T0212358

Ergonomic Insights

This book provides a great collection of work design testimonies with transferable lessons across many industry sectors and domains. It discusses physiological and cognitive parameters, teamwork, social aspects, and organisational and broader factors that influence work design initiatives.

It is important to learn from practitioner stories and real-world conditions that affect the theoretical applications of work design. Readers will benefit from understanding the struggles and successes of the authors. The chapters cover a wide spectrum of human factors and user needs, including decision-making in (ab)normal and safety-critical situations, physical ergonomics, design-in-use modifications, and tailored training. The text examines holistic approaches that lead to improved work methods, worker engagement, and effective system-wide interventions.

Ergonomic Insights: Successes and Failures of Work Design is primarily written for professionals and graduate students in the fields of ergonomics, human factors, and occupational health and safety. Educators will also benefit from using these case studies in class lessons.

Workplace Insights: Real-World Health, Safety, Well-being, and Human Performance Cases

Series Editors:
Nektarios Karanikas and Sara Pazell

The aim of the series is to host and disseminate real-world case studies at workplaces with a focus on balancing technical information with honest insights and reflections. Further, the application of a work design framework will propel this series into the literary crossover of traditional occupational health, safety, well-being, human factors engineering, or organisational sciences, into a design realm like no other series has done. Each book in this series will include cases sharing the tools and approaches applied per the work (re)design stages (Discovery, Design, Realisation). They will inform the readers with a complete picture and comprehensive understanding of the what's and why's of successful and "failed" attempts to improve the work health, safety, well-being, and performance within organisations.

Safety Insights
Success and Failure Stories of Practitioners
Edited by Nektarios Karanikas and Maria Mikela Chatzimichailidou

Ergonomic Insights
Successes and Failures of Work Design
Edited by Nektarios Karanikas and Sara Pazell

For more information on this series, please visit: https://www.routledge.com/Workplace-Insights/book-series/CRCWIRWHSWHPC

Ergonomic Insights
Successes and Failures of Work Design

Edited by
Nektarios Karanikas and Sara Pazell

CRC Press
Taylor & Francis Group
Boca Raton London New York

CRC Press is an imprint of the
Taylor & Francis Group, an **informa** business

First edition published 2023
by CRC Press
6000 Broken Sound Parkway NW, Suite 300, Boca Raton, FL 33487-2742

and by CRC Press
4 Park Square, Milton Park, Abingdon, Oxon, OX14 4RN

CRC Press is an imprint of Taylor & Francis Group, LLC

Library of Congress Cataloging-in-Publication Data
Names: Karanikas, Nektarios, editor. | Pazell, Sara, editor.
Title: Ergonomic insights : successes and failures of work design / edited by Nektarios Karanikas and Sara Pazell.
Description: First edition. | Boca Raton, FL : CRC Press, 2023. | Series: Workplace insights: real-world health, safety, wellbeing and human performance cases | Includes bibliographical references and index.
Identifiers: LCCN 2022032310 (print) | LCCN 2022032311 (ebook) | ISBN 9781032210322 (pbk) | ISBN 9781032394930 (hbk) | ISBN 9781003349976 (ebk)
Subjects: LCSH: Work design. | Human engineering.
Classification: LCC T60.8 .E74 2023 (print) | LCC T60.8 (ebook) | DDC 620.8/2–dc23/eng/20221006
LC record available at https://lccn.loc.gov/2022032310
LC ebook record available at https://lccn.loc.gov/2022032311

ISBN: 978-1-032-39493-0 (hbk)
ISBN: 978-1-032-21032-2 (pbk)
ISBN: 978-1-003-34997-6 (ebk)

DOI: 10.1201/9781003349976

Typeset in Times
by codeMantra

This edited book collection is dedicated to the families, friends, colleagues, teachers, researchers, clients, consumers, regulators, and everyone else who have directly and indirectly supported and influenced the chapters' authors in their journeys while navigating work complexities, celebrating successes, and reflecting on shortcomings.

Contents

Preface

Nektarios Karanikas

Sara Pazell

A FEW WORDS FROM NEKTARIOS

It was not long after the publication of the first edited book Safety Insights (Karanikas and Chatzimichailidou, 2020) that Sara, who kindly read it, approached me with the idea to work together on a new edited book on Ergonomic Insights, the one you are now holding in your hands or reading on your display. Although I had decided to continue working on two monographs that I have been trying to put together for quite some time, I could not resist the idea. Human Factors/Ergonomics (HFE) is an area to which I feel very close. Not only because I studied this discipline at a master's level and since then has changed how I view the world. It is because it applies to all aspects of human activities and interactions with our natural and socio-technical environments, whether this is our workplaces, social and public activities, house chores, etc. However, the scope of this book is the work environment.

Sara's suggestion to invite professionals and practitioners, HFE experts or not, from our network and beyond to share their experiences, especially from work design, greatly appealed to me. Open invitations were also the case with the Safety Insights book, but this time the scope was wider. HFE and work design do not, or should not, focus on safety alone. Work encompasses much more, and it is far broader than considering (un)safe interactions and exposures. It is about human performance and well-being at individual and collective levels while we generate deliverables and offer services safely, with quality, efficiency, timely, and other important parameters. It is equally about leveraging opportunities and managing negative risks, celebrating successes, and reflecting and acting on failures.

As such, this project was born in mid-2021. Apart from the invitations we sent to our network, sharing the project idea through LinkedIn and the Human Factors and Ergonomics Society of Australia (HFESA), who kindly supported our initiative, attracted more interest than what we had anticipated. Even more exciting was that several of the chapter proposals came from multiple authors, different industries, various regions, and diverse HFE applications.

To add some more enthusiasm, Taylor & Francis, after we agreed on this project, invited us to launch a book series entitled Workplace Insights.[1] The Safety Insights and Ergonomic Insights books are now nested under this series, and Sara and I wholeheartedly welcome interesting proposals for monographs or edited books for the same series. Sara has already started working with another colleague on an edited book for Healthcare Insights! Your proposal can be about specific industry sectors or work-related disciplines such as occupational hygiene and occupational

---

[1] https://www.routledge.com/Workplace-Insights/book-series/CRCWIRWHSWHPC

health. I hope this book will inspire you enough to set sails for your own editing and authoring adventure in our series.

After a long journey together with the 24 other contributors, whom I deeply thank for their patience with me, we are proud to present you with a great collection of 21 engaging chapters, some of them more "reflective" and others a bit more "technical". Although you might feel inclined to focus on chapters closer to the industry/discipline of interest, we invite you to consider all chapters with the same curiosity as most, if not all, of the cases presented offer insights and lessons that can be cross-transferred, provided that their "translation" accounts for the targeted context (Karanikas et al., 2022).

However, what is the big picture emerging from this book? This is what I asked myself while writing those lines. I decided to employ technology and use my very basic skills with the NVivo and Leximancer software to gain a grasp of the main themes and their connections. As you can observe from the word cloud (Figure 0.1), work and design appear most often, as expected. However, the terms systems, used, required, and management emerged with about the same frequency. Something like "Work design is required to manage and use systems [successfully]"; this is how I read it, but feel free to interpret it differently.

Interestingly, the next more frequently appearing words seem like suggesting the principal parameters related to work design (e.g. process, operations, risks, safety, issues) or necessary to design work (e.g. support, workers, change, training,

FIGURE 0.1   Word cloud generated with Leximancer from all chapters.

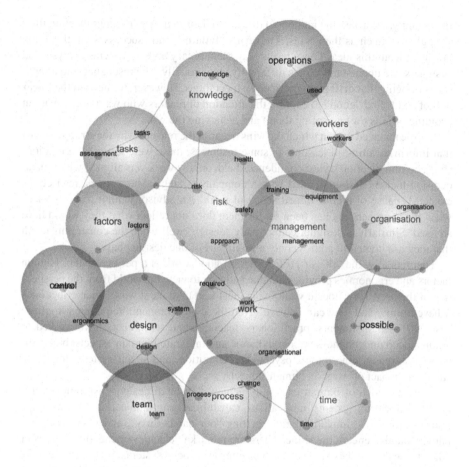

FIGURE 0.2   Concept map generated with NVivo from all chapters.

organisation, good). The outer layers of words in the cloud can be viewed as ingredients, preconditions, and contextual factors. The connections of all these terms are illustrated more clearly on the concept map (Figure 0.2), which presents several paths and interdependencies. I will refrain from expressing here what I see on this map. I will leave this to you. I hope you can get back to these figures after reading each chapter and check whether they make sense. Indeed, the words cloud and concept map do not mean to give birth to new theories and replace existing ones. They just represent the emerging picture of what 26 authors in this book have experienced and decided to share with you!

A FEW WORDS FROM SARA

Storytelling is at the heart of learning and connecting with others. We are pleased to present a collection of stories that reflect real-world scenarios of practitioners devoted to the design of good work through human factors and ergonomics (HFE).

Importantly, we asked the authors to dismantle their experiences and reflect on their struggles as much as their wins. The terms "failures" and "successes" in this book title stem from this idea. These terms may seem binary and, as systems analysis and designers, we understand things differently, dynamically, like resonance and altered states of being that drift from or exceed expectations. However, we needed the linear pull of "bad" and "good" to capture the attention of readers who want to learn about avoiding pitfalls (or navigating them) and achieving the wins.

It is the nuanced way in which the wins occur or the reasons why the struggles exist that inform meaningful lessons. In some instances, these were founded in relationships, communication styles, confidences held, or how work was aligned with crucial decision-makers and embedded within company strategy. Many times, restrictions existed because of a constrained scope of work. An advisor cannot truly "advise" when the approach that has been set for them does not reflect what they would have advised! Similarly, a designer cannot truly "design" well when the constraints mean that they are invited to address only a small part of one design phase.

Theoretically, the "fit" of HFE to design good work is a perfect match. Human factors and ergonomics provide a unique pathway to view the world from the perspective of the user: How does it work, what must I do to make it work, what effect does it have on me, and how can I thrive given these environmental/equipment/tooling/ systems? Skilled students, practitioners, teachers, and researchers have worked to attain a deep understanding about dynamic and integrated human aspects: biochemistry, physiology, kinesiology, psychology, cognition, and social systems to understand the impact of work on human subjects.

I love the idea of human capability achieved through design: inspirations, creativity, interacting systems, motivations, capabilities, communication, ideation, creations, resolutions, implementation, getting "our hands dirty", and yes, sometimes battle wounds beneath the shine. There are remarkable achievements that can result when these approaches are used such as improved social systems, better health, safer work, improved communication, environmental protections, economic advancement, and productivity with the accomplishment of serious competitive market realities. "Good work" must be designed well.

However, design can be clumsy and messy, and that sphere of uncertainty is not always comforting. It requires empathy, and that takes patience, which is not readily afforded when work requires rapid change. Sometimes, progress must be made through small acts, baby steps, to help develop tacit knowledge about the methods and the benefits. We ask business managers to bet on what is not known, and trust in a design process. This can mean navigating unchartered territories and creating unstable footing. Yet, we *know* that innovation is possible, given the right conditions.

Perhaps, through these stories, our readers will be encouraged by the successes and empathise with the authors because of the challenges they encountered, knowing that clumsiness is part of the humility of living, learning, and growing. We know that with each struggle, the only way is up, and up was the direction travelled by all chapter authors. Their perseverance is what captivates me most. The authors were challenged to walk through their experiences and tell their stories. As much as all want to sing the praises of human factors and ergonomics, they agreed to share a balanced view of their work: the roller coaster highs, lows, and side manoeuvres. This was brave of them.

Nektarios and I wanted the rawness and richness of the "day in the life of the practitioner" to be expressed to the readers as a learning opportunity. Sometimes, heavy in academia or technical jargon, there is a disconnect about how to apply what is esoteric because it is embedded in theoretical models, and blend this into the daily grind of the working world across any industry. The authors have managed to paint pictures and describe their activities in diverse industries and places, like military, aviation, transportation, agriculture, mining, oil and gas, sport and recreation, construction, nuclear settings, control rooms, education, health care, office administration, and executive management. Bravo.

I thank my colleague and the lead editor on this book, Dr Nektarios Karanikas. I have known Nektarios since he came to Australia and started teaching at his university. He strives to make his mark, create a ripple, and foster change. Notably, among other achievements, Nektarios started this literary series with the Safety Insights book and, because I commended him on telling accessible practitioner stories, then challenged him to consider the realm of opportunities with other topics that all affect the workplace and its design, he urged me to join him in this book. I am thankful for the learning opportunity. Taylor & Francis spurned this on and encouraged Nektarios and me to support a series. We look forward to other books on the horizon but, for now, we celebrate the accomplishments of this one: A celebration of human factors and ergonomics through the telling of the stories of the people that make it come alive.

REFERENCES

Karanikas, N., Shanchita, R. K., Baker, P. R. A., & Pilbeam, C. (2022). Designing safety interventions for specific contexts: results from a literature review. *Safety Science*, 156. DOI: 10.1016/j.ssci.2022.105906

Karanikas, N., & Chatzimichailidou, M. M. (2020). *Safety Insights: Success & Failure Stories of Practitioners (edited book)*. Routledge: Boca Raton. ISBN 978-0-367–44572-0, DOI: 10.4324/9781003010777

Editors

Dr Nektarios Karanikas is an Associate Professor of Health, Safety, and Environment at the School of Public Health and Social Work, Faculty of Health, QUT. He was awarded his doctorate in Safety and Quality Management from Middlesex University (United Kingdom) and his MSc in Human Factors and Safety Assessment in Aeronautics from Cranfield University (United Kingdom). He holds engineering, health and safety, human factors, and project management professional credentials and has been an active member of various prestigious international and regional associations.

Dr Karanikas graduated from the Hellenic Air Force Academy as an aeronautical engineer and worked as an officer in the Hellenic Air Force for more than 18 years before he resigned at the rank of Lt. Colonel in 2014. While in the Air Force, he served in various positions related to maintenance and quality management and accident prevention and investigations, and he was a lecturer and instructor for safety and human factors courses. After his resignation, he started his full-time academic career as an Associate Professor of Safety and Human Factors at the Amsterdam University of Applied Sciences between 2014 and 2019.

Dr Karanikas has published numerous academic journal articles, including papers in top-tier journals, peer-reviewed conference papers, and chapters, and has been invited to speak at several international and regional summits and workshops. In 2020, he published a co-edited book titled "Safety Insights: Success and Failure Stories of Practitioners". He is an Associate Editor for Safety Science and a member of editorial boards and a regular reviewer of safety and human factor-related journals. He also volunteers in various activities of professional bodies in Australia and internationally.

Dr Sara Pazell is a work design strategist and the managing director for ViVA health at work (work design specialty consultancy), working across all industries. This has operated for more than 17 years in Australia, helping clients solve real-world challenges and design "good work". Sara was awarded her doctorate in Human Factors and Ergonomics from the University of Queensland through the Sustainable Minerals Institute, her Master of Business Administration with a major in International Business Development from the University of La Verne, California, and her Bachelor of Occupational Therapy from the University of South Australia. Sara has managed business development projects in Australia, Southeast Asia, and the United States, including assuming roles as an administrator and executive director of healthcare organisations.

Sara holds affiliations with five Australian universities, including that as an Industry Fellow with the Sustainable Minerals Institute at the University of Queensland. Sara provides teaching and research support in organisational science, business management, human factors, ergonomics, health and wellness, safety, and allied health. Sara is part of the international advisory committee for the WELL

Movement concept v2 and an expert faculty member for Australia's only certified Wellness Wise™ Practitioner training programme. Sara was the committee chair for the Human Factors and Ergonomics Society of Australia's Good Work Design position paper and supporting resources. Her other passions include instructing yoga, sports, and strength and conditioning.

Contributors

Reuben Delamore
Tactix Group
Sydney, Australia

Paulo Gomes
Segurança Diferente
Brisbane, Australia

Claire Greaves
Tactix Group
Sydney, Australia

Wenqi Han
Freelance consultant
Singapore

Maureen E. Hassall
University of Queensland
Brisbane, Australia

Deike Heßler
Hochschule Osnabrück
Osnabrück, Germany

Thomas Hofmann
Hochschule Osnabrück
Osnabrück, Germany

Bharati Jajoo
Body Dynamics
Bengaluru, India

Keith Johnson
Fulton Hogan
Eight Mile Plains, Australia

Stasinos Karampatsos
Hellenic Air Force
Cholargos, Greece

Nektarios Karanikas
Queensland University of Technology
Brisbane, Australia

Svenja Knothe
Hochschule Osnabrück
Osnabrück, Germany

Alicia Lampe
Hochschule Osnabrück
Osnabrück, Germany

Christopher M. Lilburne
University of Queensland
Brisbane, Australia

Jennifer Long
Certified Professional Ergonomist
Sydney, Australia

Philip Meyer
Freelance consultant
Victoria, Australia

Graham Miller
Humans Being At Work
Brisbane, Australia

Lidiane Narimoto
EWI Works Inc.
Edmonton, Canada

Anjum Naweed
CQUniversity Australia
Adelaide, Australia

Valerie O'Keeffe
Flinders University
Adelaide, Australia

Sara Pazell
ViVA health at work
Sunshine Coast, Australia

Ruud Pikaar
ErgoS Human Factors Engineering
Enschede, Netherlands

Jose Sanchez-Alarcos Ballesteros
Factor Humano
Valladolid, Spain

Kym Siddons
Kym Siddons Physio
Adelaide, Australia

Rwth Stuckey
La Trobe University
Melbourne, Australia

Brian Thoroman
Queensland Rail
Brisbane, Australia

The contributors would like to note: The cases and perspectives shared in the chapters of this edited volume do not necessarily originate and/or reflect the authors' current employer and any other organisation, committee, or group with which the authors might be affiliated.

1 Human Impacts on Work Design

Graham Miller
Humans Being At Work

CONTENTS

Organisations are as complex as the people in them. I have been working in, consulting, and studying organisations for decades, and I'm still often mystified by how organisations function and which ingredients, when mixed, deliver effective and successful organisational outcomes – or not. No doubt there could be many reasons for this, but one would be that organisations are both 'rational' and 'non-rational'. This alleged dichotomy, in part, underpins my 'success' and 'failure' stories below and, therefore, deserves some explanation.

I suspect that many of us see ourselves, and by extension our organisations, as 'rational' entities where plans are developed, research is undertaken, information is assessed, decisions are made based on logic and data, strategies are implemented, and performance is tracked. This perspective reflects what we describe as the 'classical' view of organisations, where centralised leadership addresses discrete issues and undertakes technical/non-political activities, and where performance is measured by comparing outcomes against intentions (Dufor & Steane, 2006).

On reflection, this rational/classical perspective is my subconsciously predominant 'default' position on how organisations work or should work. As a student of Total Quality Management in the 1990s, I developed a strong affinity for 'processes' by documenting, monitoring, and improving them. The Plan-Do-Check-Act (PDCA)

DOI: 10.1201/9781003349976-1

1

model[1] and the 'process determines outcome' mantra made absolute sense to me, and still do. Although this rational/classical perspective of organisations feels 'right' and comfortable to me, I realise that, like everyone, I have biases and blind spots and recognise that there is probably more to the story. After all, organisations are made up of humans, some of whom make important organisational decisions one minute and stockpile toilet paper during pandemics the next! Marketing professionals tell us we humans mostly decide based on emotions, although often disguised as logic, rather than reasoning. Surely, this must play out in organisational life, mustn't it?

Counter to the rational/classical view of organisations, Mats Alvesson and Andre Spicer suggest that organisations are, indeed, mostly 'non-rational'. Their evocatively titled 'Stupidity-Based Theory of Organizations' (Alvesson & Spicer, 2012) claims that most managerial practices are adopted based on faulty reasoning, alleged wisdom, and a complete lack of evidence. A core of this theory is an assertion that although contemporary knowledge-based organisations claim that their 'intellectual assets' are their greatest resource, non-rational thinking, including power and domination, restricts the 'intelligent mobilization of cognitive capacities'. This is because of organisational norms; we avoid asking tough questions and searching for rational answers so as not to embarrass or threaten managers. As a result, dialogue is discouraged, justifications for decisions are not requested or provided, and 'conformity' is rewarded.

The situation above, combined with organisational time constraints, breeds a lack of curiosity and closed-mindedness and minimises critical reflection. Consequently, intelligent people working in organisations refrain from using their cognitive and 'reflexive' capacity and engage in what the authors describe as 'stupidity self-management'. The latter means putting aside doubts and reflexive concerns to minimise dissonance and, instead, focusing on 'identity construction' and career progression. In doing so, organisations undermine the latent knowledge and intellectual capacity that resides within. As a result, poor decision-making prevails and critical issues, including sanctioned immorality and questionable ethics, go unchallenged.

When I first heard of the Stupidity-Based Theory of Organizations, my rational perspective-loving self was sceptical. However, recent Royal Commissions in Australia provide technicolour examples of the very issues that the theory highlights. During Australia's Banking Royal Commission, an exchange between the Commission's senior counsel and the Commonwealth Bank's new Chief Executive Officer (CEO) confirmed that the Commonwealth Bank's auditing department had advised the Board in 2015 and 2016 that the bank's ATMs were breaching anti-money-laundering and anti-terrorism laws. Despite this being a 'red rating', they did nothing about it. In the same exchange, it was revealed that the new CEO had challenged his boss at that time (i.e. the former CEO), about selling customers 'junk insurance' when the bank knew about scandals and repayments for similar products in the United Kingdom. The back-then boss advised the now-new CEO to 'temper his sense of justice' (Ziffer, 2018).

So, is perhaps my rational perspective of organisations a little naïve? Perhaps organisations are a blend of rational and non-rational elements, and I have a bias

---

[1] https://asq.org/quality-resources/pdca-cycle

towards adopting a predominantly rational perspective of organisations. If this is the case, what does this mean for my approach to work design? Reflecting on my success and failure stories below may shed light on this.

MASKED WORK DESIGN ISSUES

Some years ago, I was working in an organisational development role for a government agency. It was a great job, working with various teams to introduce significant change programmes, facilitating planning workshops, managing the quality programme, training project teams to implement organisational process improvements, and implementing self-managed work teams.

One afternoon, a senior manager contacted me and invited me to discuss a problem they were having with one of their teams. They explained to me that the reception team, a small team of four people, was experiencing personality conflicts that had been brewing for some time. The senior manager gave me a brief history lesson on the team and its members, and a run-down on what they thought was going on, and who they suspected was the key protagonist. The senior manager asked me to work with the team to resolve the issue. I left that meeting with the impression that the senior manager expected that the solution to this apparent personality conflict may involve relocating the suspected protagonist.

I subsequently met with the team and, while maintaining confidentiality, told them as honestly as I could that I was meeting with them to seek their feedback on what they saw as the issues and help them resolve these. I also asked what their assumptions were about why I was meeting with them. From their responses, it appeared that my 'arrival on their doorstep' had not been effectively communicated, and their key assumption was that the senior manager appointed me to investigate them and report my findings back. I reassured the team members that my primary intention was to help them identify and address issues to help improve their life at work if they were willing to do this.

All four members agreed that there were issues. After some initial hesitancy, all agreed to engage in a discussion to reach a resolution. I asked each of the team members to explain the situation and the issues from their perspective and sought agreement from all in the group to let their colleagues speak, uninterrupted, and to 'just listen', which they agreed to do. While each member of the group spoke, I recorded the discussions on a mind map on a whiteboard. This visual representation of their conversation helped to 'paint a picture' of their situation and the perceived issues influencing their situation and provided a focal point for the conversation.

The discussion exemplified the value of mind mapping. The visual focus of participants is on the mind map rather than solely on each other or the floor in particularly uncomfortable situations seems to encourage more open dialogue. Translating the spoken word into a visual representation helps tie the various conversation 'threads' together, providing clarity to what might otherwise appear to be a rambling conversation. The visual representation of the spoken word also prompts inquiring questions to clarify potential misinterpretations. As the map develops, themes and key issues become more obvious. And asking 'what are you seeing here?' can spark a whole new round of conversation and potential collective insight.

At times, the discussions were passionate, but team members showed a willingness to engage meaningfully to resolve the issues. Having an independent facilitator to help guide the discussions was, I believe, critical because this provided team members with a non-threatening and impartial listener with whom they could make eye contact while still communicating with their colleagues. I paraphrased their comments and asked questions to help clarify communications among the team members. After everyone had spoken, I asked the group what they were seeing on the whiteboard. What were their common themes or connections? What were the key issues emerging? From there, I asked the group about how these issues could be best addressed.

These discussions confirmed that the team of four worked together each day to manage the reception function (triaging clients, processing applications, etc.). Although the team had a supervisor, the team was essentially self-managing. The supervisor was geographically separated from the team and supervised a larger team that was undergoing a significant systems' change and consumed much of their time. The reception team was a fresh addition to the supervisor's responsibilities, because of a recent organisational restructuring, and they had largely left the team to their own devices.

The reception team members were technically competent and experienced and could perform any of the required roles. This provided advantages of redundancy, where all roles could be covered during staff lunch breaks or absences. However, there was no job role roster. Which team member performed which role each day was determined by who placed their bags and belongings on which chair when they first arrived at work. This meant that whoever arrived earliest got more choice over which roles they performed.

Team members confirmed that, in the past, there were occasional discussions regarding who would perform which role that day. However, for unstated reasons, these often had an uncomfortable undertone, and more recently, they rarely discussed who was to perform which role. One team member admitted deliberately trying to get to work earlier than others to have a broader choice of job roles that day. This simple act highlights the importance of choice and self-determination in workplaces. Daniel Pink posits that *autonomy* is one of three key work motivators: people want to have control over their work. The other two are *mastery* (i.e. people want to get better at what they do) and *purpose* (i.e. people want to be part of something bigger than themselves) (Pink, 2009).

These discussions helped to 'clear the air', and team members agreed that the confusion around team roles and responsibilities, together with some untested and incorrect assumptions about team members' motivations, was a major contributor to the recent deterioration in team relationships. The team agreed to develop a weekly rotating roster where each person undertook a role for a week. The members also agreed to support each other when work demands varied for each of the roles. These simple structural changes co-designed by the team members addressed the aggravation which had built up, and the team environment improved from that point forward.

This was a 'success' story because of several reasons. First, all team members shared a willingness to resolve their issues. Without this shared commitment, any attempts by me would likely have been fruitless. Second, while acknowledging the senior managers' perspective on what they saw as the issues, I approached the task with an open mind and tried to avoid preconceptions of issues and solutions. Adopting a 'beginner's mind' promotes a broader perspective. Third, I was as transparent and authentic as I could be with the team members, which demonstrated respect and helped to establish early rapport and trust; this set the scene for them to do likewise. Fourth, I genuinely listened to team members to develop an understanding of the issues without judgement, which created an environment of honest disclosure.

Finally, team members developed their own solutions, which engendered their commitment to implementing the solutions. The need for self-determination in work is well-documented and long-known by behavioural scientists. McGregor's Theory X and Theory Y model, which was developed over 60 years ago, recognises that 'the essential task of management is to arrange conditions so that people can achieve their own goals best by directing efforts towards organisational resources' (McGregor, 1960). Similarly, Chris Argyris has long argued that humans have basic 'self-actualising trends', akin to plants seeking to reach their biological potential (Argyris, 1962). The implications of this for job design are significant, and the importance of co-design in human-centred workplaces is self-evident. As world-renowned facilitator Roger Schwarz acknowledges, 'When people are involved in decision-making, they have greater commitment' (Schwarz, 2017).

This experience taught me that poor work design can masquerade as something else. In this case, a lack of 'rational' workplace structures in the form of no clear role allocations resulted in ambiguities and led to a 'non-rational' response of poor team communication resulting in workplace tension, which was attributed to 'personality conflicts'.

UNDERESTIMATING THE POLITICS

As a management consultant, I was engaged by a government agency to design an 'integrated operations' approach to better incorporate its key operations. Consultants are often engaged when an organisational stalemate is reached, and, although I was unaware at the time, this was the situation in this case. The organisation had contemplated developing a more integrated approach to several of its operations for some years. However, previous attempts to better integrate 'like' operations had struggled to gain sufficient executive-level agreement and commitment.

Six months prior to my engagement, the organisation had established an integrated operations steering committee comprising the organisation's senior leaders. The committee had met several times to discuss integration concepts and options and had identified which parts and processes of its operations they could reorganise to enable a more integrated approach to some of its key business. At some point, the steering committee concluded that external help was required, and they scoped the terms of my consulting engagement. As part of my engagement, I was to report progress monthly to the steering committee on each of the identified process areas. The client described the project as 'process consolidation', but I saw it as an organisational change project. Although the client never described this as a 'work design' assignment, it very much was.

Soon after my engagement, I recognised an inconsistent use of language. Some senior leaders used the term 'coordinated operations' rather than 'integrated operations', and these terminology inconsistencies indicated to me potential differ-ences in understanding. In addition, there appeared to be an incoherent understand-ing of what the integrated operations 'end game' looked like. This was surprising, noting that they had considered the concept for several years prior to the project commencing, and they had established the integrated operations steering committee for six months.

Consequently, an early priority for me was to resolve this apparent mismatch of views and understanding. I developed a concept of what 'integrated operations' might look like, with definitions and potential pathway options showing the potential phases of implementation, from partial to fully integrated operations. This was based on an amalgam of views that had been provided to me through discussions, blended into a synthesised, tangible concept proposal for discussion. I presented this concept document to the steering committee with an expectation that this would generate discussion, expose any differing views, and build a common understanding among the group of what integrated operations could look like in reality. The steering com-mittee's discussion on my proposed concept and road map was underwhelming. Although there was surprisingly little debate, I left the meeting with tacit approval to proceed to 'Phase 1'. Ahh, success!

A large part of this project involved me facilitating stakeholder workshops, including process redesign scoping workshops with relevant stakeholders to identify and validate the processes identified by the steering committee. Through these work-shops, two of the identified processes were deemed unsuitable for a redesign. The steering committee subsequently accepted these recommendations. Partway through this, a key senior executive retired unexpectedly for family reasons. This provided a catalyst for a subsequent amalgamation of two divisions into one, which provided a welcomed revised structural platform for a more integrated approach to key parts of business operations. More success!

I continued working with various stakeholders to scope the process changes required to make integrated operations a reality. I formed process redesign proj-ect teams for the identified projects comprising a cross section of mid-level manag-ers from across the organisation. Over a two-month period, I facilitated redesign workshops to identify current process arrangements, responsibilities and outputs, and preferred future options. These operational-level workshops and meetings were productive, with team members showing collaboration and genuine commitment to adopting a more integrated approach. Constructive solutions to complex problems were created, and implementation plans were agreed upon and developed. Even more success!

The project also identified several complicated issues which needed to be addressed at an organisational level, including changes required to the legacy IT systems and changes to staffing categories and entitlements outlined in current enter-prise agreements. Upcoming enterprise agreement negotiations provided an opportu-nity for the organisation to incorporate new integrated operations requirements into the new agreements.

Nevertheless, although the project was progressing well at the operational level, I found it difficult to engage meaningfully with executive-level stakeholders. Steering committee discussions were often superficial. These meetings were one-hour duration, held monthly, and chaired by the deputy CEO. They were relatively formal affairs, and discussions were 'respectful'. Committee members were busy executives, predominantly males, and therefore with limited diversity, who were also engaged in other significant concurrent organisational change projects.

Whether committee members had read the information I was tabling at the committee meetings was unclear. Even though much of the material I presented to the committee was, at least in my mind, pivotal and important, discussions were generally brief and dispassionate. Questions were few. The only female committee member occasionally asked insightful and interesting questions, which helped to add some spark to the discussions. However, to me at least, she appeared to be a lone voice of interest and concern. Often, other committee members did not engage in conversations. Meeting time constraints meant that discussions on key issues were often left unresolved, which I had to resolve later with the project sponsor.

The project continued, and in consultation with relevant stakeholders, the project produced redesigned processes with revised accountabilities, new policies and procedures, and implementation plans for each of the revised processes. This provided a blueprint for how the organisation would implement an integrated operations model via a staged approach. Success continues!

During this development work, personnel changes meant that my reporting lines changed, and my key contact became the person who was earmarked to oversee the implementation of the new arrangements I had developed under the design phase. I found this person difficult to engage; it was hard to find time to discuss issues. There was a lack of feedback on key documents. The person was speaking in riddles and demonstrated differing interpretations of the design concepts. As long-term staff members who had progressed through the management levels, senior managers were technically oriented. They did not convince me that they were sufficiently committed to the agreed way forward or had the appropriate temperament to manage the human resource issues required for the successful implementation of an organisational change of this magnitude. Success diminishing!

In addition, it became increasingly apparent that the commitment of some senior executives began to waiver on the details of some of the more challenging elements, including negotiating with unions the proposed changes to employee classifications and roles in the new enterprise agreement. I sensed that there were clouds developing on the work design implementation horizon. Success diminishing further!

After my engagement ended, the organisation recruited a full-time project manager to help manage the implementation of the initiative of the integrated operations. This appointment coincided with the departure of the previous project sponsor. I had no conversations or handover discussions with the incoming implementation manager. Later, informal reports indicated the organisation has made some progress with the implementation. However, several of the details developed under the design project were amended, some elements of the implementation were curtailed, and several key people in the organisation had moved on. Success becomes failure!

As Schwarz reminds us 'The finish line is not when you and the group have made a decision; it's when the decision has been implemented effectively' (Schwarz, 2017, p. 82). Yes, the design component of the project provided the required 'rational' outcomes, but these were not enough to ensure that the project delivered what was intended. There are always multiple contributors to project failure or under-performance. In this case, the broader organisational issues contributed to this sub-optimal outcome, including the 'shifting sands' such as key personnel changes, and political issues that seemed to play out at the senior level. However, I was also a contributor because I failed to acknowledge or address the non-rational influences and political issues playing out at the senior level.

While I developed a proposed concept of integrated operations which the steering committee endorsed, key learning from this is that endorsement does not equal commitment. This issue required a dedicated discussion rather than incorporation into an already busy committee meeting agenda, hoping it got 'a tick'. The apparent lacklustre level of support from steering committee members should have been a red flag for me, but it was not. Instead, my ego allowed me to believe that I had developed a solution to an apparently intractable organisational problem, and the steering committee's endorsement of this solution meant that my worth was proven!

I worked closely with mid-level managers as part of the project; however, my engagement with the senior leaders was predominantly through monthly steering committee meetings. This approach did not engender genuine commitment from steering committee members. The design solution I proposed was mine, not theirs. As with all significant work design projects, senior executive commitment is critical. Gaining genuine executive-level support for the project would, at the very least, have required more robust discussions to understand the thinking of the committee members and tease out an agreed position that all senior executives supported. Alternatively, I could have engaged the committee in a facilitated discussion to develop their own agreed position on the 'integrated operations' solution. I believe this would have increased the chances of organisational success. I could also have engaged more meaningfully with senior leaders at critical points during the project, for example, through one-to-one meetings, to explore their concerns or perceived issues.

On reflection, I attribute my contribution to the project's failure to not acknowledging or addressing the non-rational issues that emerged. The question is why was that? Writing this chapter prompted me to reflect on this, and I think perception is the answer. Perception matters because the way we perceive organisations will determine what we see. What we see as important is what we gravitate towards.

WHY WAS THAT?

A model that has helped me make sense of the 'why was that' question is Bolman and Deal's 'four frames' model (Bolman & Deal, 2021). This model identifies four lenses, or 'frames', through which we can view organisations:

- a *structural* frame, characterised by goals, tasks, roles and responsibilities, metrics, etc. This broadly aligns with the rational/classical perspective of organisations mentioned above.
- a *human resource* frame, characterised by employee needs, professional development, job satisfaction, etc.
- a *political* frame, characterised by personal agendas, power plays, coalitions, etc.

- a *symbolic* frame, characterised by motivating staff, visioning, celebrating victories, creating purpose and meaning, etc.

These four frames provide a broad, holistic perspective of organisations. Although this model was developed to inform organisational change programmes, it helps to explain the rational/non-rational dichotomy, where some frames reflect a rational orientation, and some reflect a non-rational orientation. On examination of the four frames, I recognised that I have a natural affinity for the *structural* frame. It is my preferred perspective for the reasons discussed above. This is probably unsurprising for someone who has spent 20 years in the military and who consults in the predominantly rational areas of planning, risk management, emergency management, and business continuity. As I see myself as a people person, I also identify closely with the *human resource* frame. I'm often described as a good listener; I enjoy hearing people's stories and I rate job satisfaction as critically important for organisations and the people in them. The symbolic frame is something I can identify with, too. I see creating purpose and meaning as important if we do it authentically and without 'hype'.

However, the *political* frame does not resonate with me; it is my least preferred frame. I have always had a mild disdain for organisational politics. I much prefer it when everyone gets along and works together harmoniously. I value transparency, diversity, and teamwork, and I see no value in power plays and personal agendas. However, I recognise that not everyone shares this view. I recall a colleague once telling me how much they loved the 'cut and thrust' of working in the minister's office, and I remember thinking, 'that's not for me'. For me, the political frame is synonymous with agitation, confrontation, and potential conflict, and I don't like confrontation or conflict.

On reflection, I attribute my aversion to confrontation and conflict to my childhood experiences. My childhood observations told me that my father was uncomfortable with confrontation and conflict, and I am too. In the household I grew up in, avoiding confrontation and conflict became a useful survival strategy that I subconsciously honed from childhood. These principles have followed me into adulthood, and now, I have a lifetime of cleverly but unconsciously developed strategies to help me avoid confrontation and conflict. I'm often described by others as amenable and easy-going. On the 'Big Five Personality Test',[2] I rank high in the agreeableness factor. These seemingly laudable traits of mine reduce the risk of confrontation and conflict and help me feel flummoxed and my brain slow down. I believe that my aversion to the political frame and its perceived association with confrontation and conflict shines some light on why my success story was a success, and why my failure story was a failure.

In my success story, the team members conflicted with each other because of a structural issue. Dealing with rational structural issues is my sweet spot. In this environment, I was in my happy place. Thankfully, the team members agreed to talk things through rationally, which I envied, and the discussions resulted in a favourable outcome. Great. This experience reinforced for me that I am an extremely competent

---

[2] https://bigfive-test.com/

group facilitator in the right circumstances. Had the team members not agreed to work through their issues rationally and had things gotten out of hand, my conflict aversion would probably have kicked in and my facilitation prowess may have deserted me. If this had happened, it is likely that this experience would not have been one of my success stories.

My failure story is even more enlightening. I approached this assignment with my predominant rational, structural perspective on high beam. There were lots of structural issues to address, and I did just that. I engaged with eager people and facilitated operational-level workshops to develop improved rational structural arrangements. This solution addressed what the organisation had been struggling with for some time. However, there were also some important non-rational political issues at play with senior members of the organisation and with the person who would implement the redesigned organisational structures and processes. My aversion to the political frame and my natural inclination to avoid confrontation and conflict meant that I underestimated these important non-rational issues. However, a more honest assessment would attribute this to the fact that I felt intimidated by the senior leaders. Challenging or confronting them in steering committees or elsewhere was not something with which I felt comfortable. I just did not want to move to this space, so I didn't. As a result, what could have been a successful work design project became a failure story.

The Lessons for Work Design and Beyond

We can all learn from our successes and failures. Some lessons link to a specific experience, and some are more broad-reaching that emerge multiple times across different circumstances. If we learn these lessons well, hopefully, our future experiences that may have otherwise ended up as additions to our failures list will end up on our successes list. Below, I share some of my broad-reaching lessons that emerged from all my success and failure stories.

Organisations Comprise Rational (Mostly Visible) and Non-Rational (Mostly Invisible) Elements

Our 'world view' of organisations will impact our awareness of the interplay between these elements, filtered through our preferences, biases, and blind spots. Being aware of these filters will enable us to adopt a more holistic and complete picture of the organisation we are working with and the work design requirements and issues. This is particularly important for work design, which could arguably be seen as inherently rational because of its scientific origins and predominant focus on structural issues.

Work Design Is Synonymous with Organisational Change

A successful work design project includes a successful design component (largely rational) and a successful implementation component (largely non-rational). The implications of this are significant to work design success. The fact that 'organisational politics can become a problem during times of organisational change' (McShane, Travaglione, & Olekalns, 2013) suggests that the successful implementation of work design initiatives will need to address both the rational and non-rational aspects of organisations. Consequently, work design practitioners need to be acutely aware of the organisational politics that could be at play if they want to see their work design projects implemented as intended.

ALL OF US BRING OUR 'FAMILY OF ORIGIN' ISSUES TO WORK

Our education in group dynamics starts at a very early age, in our family unit. This is our first team experience. Our young and impressionable brains learn quickly how teams and groups work and where we fit in them. This knowledge is later complemented by lessons from the schoolyard. Through these experiences, we learn, for example, how to deal with conflict, how to respond to compliments or criticism, and how to persuade others to achieve our desired outcomes. For many of us, these childhood lessons may not have reflected best practice, noting that they originated from the courtesy of role models who were operating in accordance with their own 'family of origin' programming. As a result, we may freeze when approached by a boss who reminds us of a critical parent or unjustifiably resent a work colleague who subconsciously reminds us of a disliked sibling. With all our 'family of origin' issues playing together in the workplace sandpit, it should be no surprise that some elements of organisations are non-rational! Developing our awareness of these hidden aspects of ourselves can only benefit our workplace interactions and beyond.

EGOS CAN JEOPARDISE SUCCESS

The proverb 'pride comes before a fall' applies as much to work design practitioners and consultants as anyone. Our ego wants us to design a brilliant solution to a problem and be lauded for our cleverness. However, successful work design solutions which are implementable and are implemented require ownership from those impacted. That means that effective solutions are as much about best fit as best practice. For work design practitioners and consultants, this requires taking people on a journey, and never losing sight that it is about them, not us. We must keep our egos 'in check'.

PERSONAL REFLECTION IS POWERFUL

The pace of organisational life allows little time for reflection. However, reflection is key to addressing each of the points above and more. '*How we spend our days is, of course, how we spend our lives*' (Hasson, 2020). If we want to live better lives, we must live better days, and reflection can help us do this. Reflection can be achieved in different ways (e.g. engaging a counsellor or an executive coach, undertaking an 'action learning' process, journaling, and meditating), but it is a deliberate process. Reflection will help uncover amazing things that are quietly floating just below the surface.

FINAL THOUGHTS

For me, writing this chapter has been a valuable reflective process. It has prompted a discussion with my partner, who happens to be an executive coach, to help me make sense of my subconscious beliefs that were formed courtesy of my 'family of origin' experiences. A key part of this has been acknowledging my aversion to confrontation and conflict and my apparent inability to manage this effectively. I have never been trained in how to do this effectively. Instead, I developed an approach based on role-modelled avoidance. The downsides to this, of course, are many. One has been a self-perception of being a 'fair-weather' facilitator. Another has been cognitive dissonance around what I did or didn't do and what I should have done. However, through reflection, I now have a better understanding and, as a result, have adopted a new mindset. This has been like turning on a light in a previously dark room. I have made significant progress in only a short time, and this, I believe, has and will continue to result in me being a more effective work design practitioner and a more effective human. I highly recommend it!

BIBLIOGRAPHY

Alvesson, M., & Spicer, A. (2012, June 21). A stupidity-based theory of organisations. *Journal of Management Studies Vol 59 number 3*.

Argyris, C. (1962). *Interpersonal Competence and Organizational Effectiveness*. Homewood: Irwin.

Bolman, L., & Deal, T. (2021). *Reframing Organisations*. Hoboken: Jossey-Bass.

Dufor, Y., & Steane, P. (2006). Competitive paradigms on strategic change. *Journal of Strategic Change Vol 15 number 3*.

Hasson, G. (2020). *Mindfulness Pocketbook*. Chichester: John Wiley & Sons, Ltd.

McGregor, D. (1960). *The Human Side of Enterprise*. New York: McGraw-Hill.

McShane, S., Travaglione, T., & Olekalns, M. (2013). *Organisational Behaviour*. Sydney: McGraw-Hill.

Pink, D. H. (2009). *Drive: The Surprising Truth About What Motivates Us*. New York: Riverhead.

Schwarz, R. (2017). *The Skilled Facilitator*. Hoboken: Jossey-Bass.

Ziffer, D. (2018, November 24). Banking royal commission exposing Australia's business leaders aren't operating on a higher plane. Retrieved from ABC News: https://www.abc.net.au/news/2018-11-24/banking-royal-commission-commonwealth-bank-bosses-not-learning/10549754

2 The Underestimated Value of Less-Than-Ideal and Proactive Ergonomic Solutions

Kym Siddons
Kym Siddons Physio

CONTENTS

I have worked in private physiotherapy clinics for most of my 25-year career as a sports and exercise physiotherapist. In mid-2015, I moved from overseas to a practice located in inner-city Adelaide, Australia. I noticed that my clients from the surrounding offices and high schools came in to see me with different challenges. Several of their physical complaints were related to conditions arising from poor work environments and study habits. I noted that desk workers and students presented a high incidence of low and mid-back pain, neck and shoulder pain, and headaches. Thus, I began my journey into researching and upskilling in occupational health physiotherapy with a focus on ergonomics.

My experience was backed by emerging research highlighting the significant health risks associated with excessive sitting and poor workstation ergonomics. Such evidence gave me hope and fuelled my determination to help spread the 'good news' about the simple yet significant steps that organisational leaders and workers could take to optimise the health of their employees and minimise the risk of ill-health and injury. This includes increasing workers' capacity for office-based work and limiting their risks of incurring musculoskeletal injuries (MSDs).

DOI: 10.1201/9781003349976-2

When working with elite athletes, our high-performance team of health professionals and coaches screened the athletes for injury risks, increasing athletic capacity (e.g. strength, power, flexibility, and skills) and managing their workloads to optimise their performance and minimise their health risks. I realised that workers and students might not have such a well-formed team of health professionals and coaches on hand. However, I believe that it is invaluable to educate and empower them in practical means to identify when they are at an increased risk of injury, ways to increase their performance capacity, and methods to manage their workloads for better health and productivity.

One means I find useful for the prevention and early intervention of musculoskeletal conditions is conducting ergonomic assessments that include a review of workers' workstations with subsequent recommendations and advice about posture, movement, and targeted exercises. From a clinician's point of view, by the time a client comes to see me in the physiotherapy clinic, it is usually when their symptoms have reached a point where their pain or dysfunction is significantly impacting their daily life. Sometimes, the pain or dysfunction has built up over time, and the client and I work through the factors that have contributed to their problem. In these cases, assessing the environment in which they work, such as their workstation set-up, is a critical factor in managing their MSDs and implementing sustainable improvements.

Other times, the client's symptoms or dysfunctions are acute. Even in this instance, together, we can usually trace back the 'warning signs' that their body wasn't coping well with the various loads it needed to sustain on a day-to-day basis. We then focus on improving their capacity to sustain these loads (e.g. via combinations of strategies such as targeted exercises and stretches, education, and manual therapies) and changes that can be made to their environment to enhance their performance. This is where optimising ergonomics within their work, while considering home and recreational environments, is of utmost importance.

In addition to considering the physical needs of a worker or student, an enquiry about their mental health is necessary (WHO, 2000). Stress, anxiety, and depression may affect them by slowing cognition and reducing job performance and productivity. This can also affect their physical capability and daily functioning (Lerner & Henke, 2008). Reviewing these factors is therefore key to conducting holistic ergonomic assessments and making tailored, appropriate recommendations that will be successfully implemented and sustained by both employees and employers.

Having read how ergonomic assessments consider both the physical and mental health of an employee or student, it might seem obvious that conducting an ergonomic assessment must be done with the person who uses equipment in their environment. Yet I cannot tell you how many times well-meaning employers or organisations have told me 'That person is away today, but can you still take a look at their desk?' As much as I aim to be obliging, the answer is 'No'. I cannot ask the user how they feel physically and mentally during and after using their workstation. Nor can I objectively assess how their body fits and moves within and around it. Ergonomics is not about evaluating inanimate objects; it is about humans, and how they interact within their work environment, and what they need to thrive.

Neglecting to assess and address the environment in which a worker or a student spends most of their day undoubtedly puts their physical and mental health at risk, can delay progress in the management of MSDs, or contribute to the recurrence of health problems. An 'office sweep' of ergonomic assessments by a trained professional is often very enlightening for employees and employers as we work together to determine good work design that benefits individuals' work outputs while enhancing their physical and mental health. In the following cases, employees opted to have their office workstations assessed.

TEMPORARY SOLUTIONS COULD WORK AS WELL!

Early in my professional journey, I performed ergonomic workstation assessments in a large office for workers who 'opted in' to have their workstations reviewed to generate improvement recommendations. These employees were offered the assessment, and their participation was voluntary. There was one gentleman I'll never forget. His complaint to management after I assessed his workstation changed my approach to my future work in ergonomics. While I was highly embarrassed at the time and keen to remedy the situation urgently, I'm eternally grateful for those lessons learned.

When I entered his office, this man in his mid-thirties was friendly and eager to see me. He went to great lengths to explain all the details of his previous injuries and current levels of discomfort. He complained of low back pain which increased by the end of the day and pain in both wrists and hands while typing. By using the Visual Analogue Scale,[1] I determined that this was low-grade pain intensity, rated at 2–3/10 typically and less than 5/10 on most occasions. He reported mild tightness in the back and forearms but minimal to no sensory changes such as pins and needles or other physical or neurological symptoms.

I could immediately see several factors about his chair, desk, and screen set-up that might have been contributing to overload on his body. Further questioning about his posture habits and how often he moved about during the workday also revealed areas for improvement. He was not particularly active but did walk or jog slowly a couple of times a week before or after work. Hence, we began the assessment, adjustment, and education process.

INITIAL ASSESSMENT AND SOLUTIONS

The client's desk was of fixed height and his workstation was arranged so that the chair and screen/keyboard set-up was in the corner of a right-angled desk. His chair had no arms but was unusually high and the seat pan was tilted forwards. While the seat pan tilt helped the balls of his feet touch the floor, the client was unable to support any weight on his buttocks or recline against the backrest during the day. This likely resulted in excess use, leading to fatigue of his back and neck muscles. It also led to improper positioning of his shoulder blades because his arms reached forwards so the shoulder blades were less effective in anchoring and providing support for his arms.

---

[1] https://www.sira.nsw.gov.au/resources-library/motor-accident-resources/publications/for-professionals/whiplash-resources/whiplash-assessment-tools/assessing-pain-vas

Therefore, along with sitting for prolonged periods at his desk without regular movement breaks, his chair position was a likely factor contributing to low back strain and resulting pain as his workday progressed. Moreover, the forward chair tilt also resulted in him leaning on his forearms, in fixed shoulder internal rotation while typing. This likely increased the work of the wrist extensor muscles and inhibited larger proximal muscle groups functioning optimally.

We adjusted his seat pan to a neutral position and lowered it slightly. We discussed the benefits of positioning his body in the rear of the seat pan so that weight could be taken through his pelvis (i.e. via 'butt bones' or ischial tuberosities) and his back against the chair's backrest, at least part of the time, to reduce strain on his back. Then, the lumbar support needed to be adjusted to adequately fit his lumbar lordosis for comfort.

Once the chair was adjusted and he was positioned adequately to allow for the relaxation of larger muscle groups, it was evident that a footrest was required. The client's feet were far from flat on the floor, so he quickly tucked them under his chair to support them on the base. This brought his weight forward from the pelvis and chair again. When I explained how this could sabotage his efforts to use his backrest if sustained, he agreed that a footrest was likely a good option.

Moving the client's workstation out of the corner position to along one side of his desk allowed his elbows to move freely, rather than leaning on them for support. Also, significant findings of an assessment of his desktop screen and accessories showed that the computer monitor screen sat low on the desk, with his lower neck and upper back flexed, the weight of his head much farther forward than his shoulders and trunk, resulting in upper cervical (neck) extension. Sustaining this position for prolonged periods likely contributed to upper back and possibly arm pain, so we repositioned the screen.

Temporary reams of paper were used to raise the monitor so that the 'working part' of his screen was approximately eye height and education was provided in alternative postures of the head and neck to reduce neck, back, eye, and arm strain. We also discussed his keyboard and mouse positions, bringing them closer to his body to avoid over-reaching. Now, his chair and desk set-up allowed resting his back, better head/neck position, and less need to lean on his arms. He reported that the keyboard and mouse felt easier to use.

Finally, we recapped on moving more regularly, ideally, every 20–30 minutes, positioning himself to reduce strain on his back, neck, and arms often, before symptoms came on each day. Additionally, I prescribed a series of targeted stretches and exercises for his low and mid-upper back, neck, arms, and hands. He practised these with me. I left him with a double-sided handout that had documented the movement, posture, and exercise strategies that he was to focus on, with some pertinent notes and reminders written down.

Twenty minutes had been allocated for each staff member's ergonomic assessment and included brief questioning about their physical and mental conditions relating to their work and workspace, along with the objective assessment, provision of some posture and movement advice, and some preliminary adjustments. While this gentleman's assessment took double that time, by the end he was calm and comfortable with his new set-up and the challenge of developing some new movement, posture, and exercise habits.

In my detailed report, I outlined the assessment results, changes made, and education provided during the session, and I stated 'future recommendations'. These recommendations included a large, sturdy, carpeted, and height-adjustable footrest plus a permanent monitor raise. Based on his low to moderate grade symptoms and the two pieces of equipment recommended, I assessed his needs as a 'moderate priority'.

TECHNICAL MISSES AND REMEDIES

At the end of the day's assessments, I chatted with the 'People and Culture Manager' to summarise a few clients with high-priority needs. There were several workers in this office with unsuitable workstations coupled with high and dysfunctional pain levels, so we mainly discussed plans of action to help them as soon as possible. Nevertheless, it seemed remiss in hindsight that I failed to draw more careful attention to the client's case described above, and I only assessed him as a 'moderate priority' for implementing recommendations. On reflection, his reported lower levels of pain pre-assessment, the changes made that reduced physical strain, and the resulting comfort at the end of the session were the main reasons that I didn't prioritise his needs as 'high priority' or urgent.

Within a week of the assessment, I received a phone call from the People and Culture Manager explaining that this client had approached her complaining of increased wrist and hand pain. I immediately emailed the client to schedule a time to review with him the soonest possible time and reassured him that we would find a solution. I attended the office the next day and spoke with the manager and the client. I asked if any of the recommendations had been implemented yet for any of their employees. I was, naively, surprised to hear that they had not.

The client had become anxious about the wrist and hand pain, even though it hadn't increased beyond his previous pain levels but hadn't improved much either. He had expected that it would disappear quite quickly. I also noticed that I hadn't lowered the tabs on his keyboard at the time of our initial assessment, which we then raised. Lowering the tabs subsequently reduced the degree of his wrist extension during typing and improved his comfort.

I also recommended that he trial a keyboard-length gel wrist support to reduce contact stress on his wrists during typing. We discussed various alternatives for mouse use (e.g. vertical, contour, or roll-bar mouse), but he was not keen to try those in the first instance due to poor results previously. Furthermore, after realising that the lack of footrest discouraged him from positioning himself back in the chair and using the backrest, I hunted down a temporary one. A sturdy box approximately the right height for him to support his feet flat increased his comfort when sitting and served as an interim footrest.

We also reviewed his stretches and exercises. Some of them were being done too vigorously and slightly incorrectly, so modification and reiteration of key points helped fine-tune their performance. While we didn't video them in this instance, I've found that using a client's smartphone for videoing exercises helps provide a more accurate reference. I spoke with the manager, and we ordered the necessary footrest and keyboard wrist support immediately. The monitor raise was deemed more of an aesthetic feature rather than of urgent functional need, so it was not considered a high priority.

The follow-up via a phone call the next week was positive. The new footrest and wrist support were in situ, and the client was happy with them. His symptoms were infrequent and less intense, so he felt that the changes were successful. He also reported being more diligent with movement breaks and doing his stretches/exercises. I provided another follow-up via email a couple of weeks later and confirmed the success of his interventions. The client felt that his set-up, healthier posture, and exercise habits were responsible for the absence of back and arm pain. He was happy for me to carry on independently with the option to contact me always open. I reported this to the manager, and we were both happy with his results.

THE CRUCIAL MISSING POINT

Despite the additional interventions that led to dramatic improvements for the client, I realised that I had not considered psychological risk factors that would likely have improved his experience post-assessment and in the interim before the 'further recommendations' were implemented. Since this assessment, I learned to take more notice of the client's psychological status and note any risk factors that warrant prioritising their follow-up.

The flags that indicate the need for follow-up can be varied but can commonly present as apparent resistance to change; overt concern or anxiety about the symptoms or injury, including providing excess details about their injuries; use of negative language or verbal and non-verbal expression about workstations, culture, or the organisation; and reporting their excessive workloads, both physically and mentally, or their conflicting demands. Health professionals often refer to such risk factors as 'red flags' as they could potentially be more serious or complex to manage. I now report these clients as 'higher risk' and recommend that they be followed up with 'high priority'.

Change management and workplace culture are also areas that I find that need to be approached with caution. Being sensitive to the climate of the workplace, the employees within it, and the openness of management to receive recommendations and act on them is also important. Examples of this might be a 'stressful' environment that covertly discourages workers from movement breaks as they are seen to be 'not working hard' or management that, for various reasons, resists spending money on adequate furniture or accessories to support their employees' well-being.

I have come to expect that it may take organisations a bit of time to implement changes, even in 'urgent situations'. Understanding this has prompted me to ensure that I talk clients through the things that they can do meanwhile, including moving regularly, stretching, and exercising, adjusting their posture, and positioning. I try to provide pertinent specifics per a handout that includes these tips and exercises along with the 'further recommendations' that I will make based on their assessment findings. In many cases, employees are not privy to their assessment reports. Hence, oral and written communication cements the information that we discuss at the time of assessment for their ongoing reference.

I ask them to keep open lines of communication with their employers and let them, or me, know if they have any questions or concerns post-assessment. I explain that while I might make recommendations in my report, I'm not responsible for acting on them and cannot guarantee when or even whether they would be implemented. I encourage them to check in with their employer about the progress of implementing any 'further recommendations' in a timely manner. Different organisations have different approval processes, ordering, and supply of recommended equipment, so the lead time for implementing change varies greatly.

Implementing as many temporary measures as possible to support the client physically within their workstation set-up is also successful to facilitate enduring workstation-use changes. This gives the client the opportunity to try various measures to see which suits them best, allowing them to be actively involved in the design process and provide feedback. It also leaves them with the support in place rather than waiting for a deferred strategy. In this case, had I provided a temporary footrest to facilitate the development of new postural habits and paid closer attention to his keyboard and lowered the tabs, the client may have experienced a reduction in symptoms quickly post-assessment.

After 25 years as a health professional, I am still constantly learning. That's one of the things, along with the thrill of helping clients feel and function at their best, which inspires me and brings me great satisfaction. Reflecting on this 'failure' prompted me to make essential changes to my approach, my assessment, and the systems of communication and follow-up that I use that have enhanced the outcomes for my clients since.

FROM GOOD TO ALMOST IDEAL

A few years ago, I came across the most creative alteration of a workstation made by an employee that I had ever seen. I am compelled to share this story because helping this employee navigate a safer and more successful outcome was challenging but also a lot of fun! When I entered this office environment, it was because I had been engaged to conduct individual ergonomic assessments for employees who opted in. Some of the staff were giggling and were in good humour. 'Wait until you see Jay's desk!',[2] they laughed and pointed to her workstation where she was standing with a big grin on her face. It appeared that she had modified her ill-fitting standing station with various boxes (Figure 2.1) so that it looked more like a 'tower of terror' to an ergonomist (Figure 2.2), but I grinned back and walked over to start our assessment.

FROM GOOD TO BETTER

Jay explained that her low back and anterior hips became very sore during her one-hour daily commute to work via car. By the time she arrived at work, sitting down for administrative work was extremely uncomfortable. Hence, she had requested a standing desk and had been provided one since. Unfortunately, the desktop standing station that was provided approximately six months prior to my visit was not height-adjustable, nor did the mechanism work to lower down to the sitting position. She had no chair assigned to her workstation and confessed that she did not sit down, except in the lunchroom during breaks. Standing all day did take its toll: She felt tight

---

[2] *A fictional name has been used to protect client confidentiality*

FIGURE 2.1   Approach view of Jay's fixed standing workstation before the ergonomic assessment.

in her legs and lower back in the afternoons. Yet, sitting had been more uncomfortable, so she just 'put up with it'. Thankfully, her work environment and role allowed her to move around often. Yet, she was often standing still at her desk for prolonged periods of up to two hours throughout her workday.

Assessing Jay within her workstation showed that the height of the keyboard tray was far too low. She had subsequently placed the keyboard on the top section of the 'standing desk' and propped one end on a box to make room (Figure 2.3). It was turned towards the right, rather than centrally between screens. Her mouse was also squeezed onto the box beside the keyboard, on a mouse pad that hung off the edges. Moreover, her screens were grossly uneven in height. The laptop was placed on the desk, so it was quite low and the monitor at its maximum height. This monitor height was appropriate but its positioning to the right of the desk resulted in an awkward standing posture, twisted to the right. On questioning, Jay answered that she used the dual screens equally. Therefore, we suddenly became aware of the repetitive neck

FIGURE 2.2 Side view of Jay's standing workstation before the ergonomic assessment.

flexion/extension and rotation, coupled with trunk rotation that had been occurring during her computer and keyboard use.

The desk was also extremely cluttered, under, on top, and around the standing desk (Figure 2.1). We cleared some items away and found a chair for her so that we could assess her seated work approaches and determine how we could create a comfortable option for her. I explained some disadvantages of standing at her desk for prolonged periods, which included those that she had experienced, such as tight calves and lower back. She agreed that being able to sit down at her desk for periods throughout the day would be ideal if she could feel comfortable.

An assessment of the available seat showed that the seat pan in this chair was not deep enough for Jay, meaning that the distal part of her thighs was unsupported. To compensate, she tended to alternate between tucking her legs under her chair, sitting on one leg crossed under the other thigh, or perching on the front of the seat.

FIGURE 2.3   Front view of Jay's fixed standing workstation before the ergonomic assessment.

Furthermore, neither the height nor lumbar support was adjustable in the backrest. Yet, it reclined sufficiently and could be locked or unlocked to allow movement.

I talked her through some posture tips that could help her comfort while sitting. These included sitting her 'butt bones' back in the seat to avoid posterior pelvic tilt for prolonged periods, which can increase loading on spinal, shoulder, hip, and pelvic complexes, and resting her back on the backrest to reduce muscle strain. This also gave her more support for her thighs and reduced hip overactivity, which can lead to strain of the low back, pelvis, and hips. We also cleared an area on the desk beside her standing station so that she could sit down throughout the day. We discussed her need for a new chair and a sit-stand workstation but, in the meantime, moving her laptop, keyboard, and mouse down onto the desk would provide her with a sitting option that felt comfortable for short periods.

At the time of the assessment, by using reams of paper, we were able to raise and centralise the keyboard, and we broadened the mouse-use area. A different box of appropriate height was used to raise the laptop screens, and both were moved closer together to facilitate ease of use between them with reduced neck and trunk movement. While these were temporary and ungainly looking measures, they provided a comfortable interim measure.

Moreover, I demonstrated a variety of standing postures that Jay could position herself in to change her weight distribution and relieve the load on her body while standing. We also practised a variety of stretches and exercises to improve her flexibility and endurance in standing. Importantly, we practised stretches and exercises in the sitting position that helped improve mobility of her hips, low back, and trunk, with others to strengthen her core stability. Sprinkling these into her workday, along with frequent posture adjustments and a more 'ergonomically friendly' workstation, would likely help to improve her work capacity and comfort during sitting.

I highlighted Jay's situation as 'high risk' because she was suffering considerable symptoms daily. While the People and Culture team had been aware of the multiple modifications that Jay had needed to make to alter her desk to be 'semi-usable', no one had really thought about the health and safety risks. In their attempt to address her complaint about painful sitting, they had provided a low-cost desk that was completely unsuitable. They had not consulted with an ergonomic or health professional in advance to ensure that the best solution was reached.

As it is often the case, based on my experience, the People and Culture management team were likely conflicted between short-term cost savings and long-term return on investment, because budget constraints were repeatedly mentioned. It is also interesting how this employee adopted an 'I'll make the best of it' situation and was unaware of the risks for her health in accepting such a compromise. This is the flipside of psychological risk factor assessment, where an individual's positive disposition and tendency to avoid 'a fuss at work' increased the worker's risk of injury.

I talked Jay through how she might be able to adapt the principles of Alan Hedge's ideal work pattern of sitting-standing-stretching[3] to suit her physical needs and workflow. Jay believed that sitting for 20 minutes, standing for eight minutes, and stretching or moving for two minutes might not allow her to focus on 'deep work' for long enough periods. Nonetheless, she agreed to try standing less of the time than she had been, sitting for more time than she had been, moving more frequently between positions, and incorporating stretches and exercises in between. This all hinged upon her receiving the new sit-stand workstation and chair.

I also discussed with Jay her sitting position in her car seat and talked about some ergonomic, movement, and posture principles that she could apply when adjusting her car seat to reduce strain on her hips and low back while driving. She thought that bringing the seat closer might be pertinent to reduce hip flexor overactivity and leg extension while improving lumbopelvic positioning. We also practised small lumbo-pelvic and trunk movements that she could easily do while driving and sitting at her desk to reduce stiffness and soreness.

FROM BETTER TO ALMOST IDEAL

A new chair with a deeper or depth-adjustable seat pan was necessary to facilitate comfortable sitting. The People and Culture team, myself, and Jay agreed on a plan that included ordering a desktop, fully height-adjustable stand-capable workstation, preferably an electric one with a memory function that would enable Jay to easily and frequently move between sitting and standing positions. Before leaving the office that day, I spoke with the People and Culture Manager about a plan to fast-track the ordering of her equipment. Their usual process for gaining approval of a request for a stand-capable desk was typically a letter from the employee's treating doctor or therapist. This had, however, failed to result in the prescription of appropriate stand-capable desks for other employees also.

---

[3] https://ergo.human.cornell.edu/CUESitStand.html

Thankfully, the employers agreed to order the chair and electric sit-stand desktop workstation that I recommended as soon as possible. Had they been hesitant to do so, it would have been an occasion where I would have felt it necessary to subtly point out their legal obligation to provide a safe and healthy work environment. On many occasions, I have seen employers, or their People and Culture team, fail to consider or facilitate the 'fit' of an employee with their work and work environment to ensure that they can work safely and effectively, as is outlined in legislation (WHS Act, 2012).

At the follow-up three weeks later, the desk and chair were in situ for this client. Jay was experiencing significantly reduced symptoms in her back and hips, and she found that working between sitting and standing was a comfortable strategy. Her commute was also easier, and she was arriving at work and home at the end of the day less stiff and more refreshed. I encouraged her to communicate any ongoing questions or concerns with her manager or me.

PROACTIVENESS PAYS OFF

Since this assessment, I have endeavoured to encourage organisations to develop systems for employees and employers to communicate more effectively and efficiently regarding potential health risks. This can be, of course, a delicate topic that requires a careful approach, because no organisation wants to feel accused of being negligent in its communication methods. I believe that it is up to both parties, the employee and employer, to foster a 'culture of care' that allows employees to voice their health concerns without fear. It is also the responsibility of the employer to approach employees when they are concerned about their health and safety and work towards solutions in a timely manner.

In this instance, it became evident that the People and Culture management team were aware of the inadequacy of Jay's standing workstation but accepted that she was 'okay with making do'. They failed to see the health and safety risks or the urgency to engage a specialist consultant. Thankfully, this annual 'office sweep' of ergonomic reviews brought Jay's needs to light and addressed them before her situation worsened. Other employees were also suffering from ill-fitting sit-stand workstations so the need for more individualised assessment and prescription of these by a trained ergonomist was suggested to management for their future consideration.

Many inappropriate stand-capable desks had to be disposed of. That was a waste of company resources that could have been avoided by the prescription of better products. It is difficult for untrained people to assess the appropriateness of a sit-stand desk because most information online is distributed by potentially biased parties with commercial interests whose priority is to sell their products. In addition, governmental and educational bodies' policies typically mention taking heed of a referral for a sit-stand desk from a 'medical professional' in the case of injury or medical condition.[4]

Many medical professionals, however, have no training in ergonomics and can be unaware of the nuances in sit-stand desk prescription. Several factors should be

---

[4] https://www.safety.uwa.edu.au/health-wellbeing/physical/ergonomics/workstation/sit-stand-desks

assessed, such as the 'fit' of a sit-stand desk for the frame of the client, their tasks and workflow, the dimensions of the workspace, and the other equipment they use. The time lost to poor health and productivity because of delayed intervention, plus the cost of purchasing inappropriate furniture, typically far outweighs the cost of engaging an ergonomist.

In this instance, no employees of this company had lodged a claim through 'Return to Work SA' relating to the pain and stiffness that they reportedly experienced throughout their workday at their desks. Many were seeking physiotherapy treatment at their own expense. Yet, this ergonomic assessment and intervention was undoubtedly timely to avoid further aggravation and cost for both this employee and employer. I constantly try to encourage employees and employers to maintain an open yet respectful dialogue about their comfort and satisfaction with their work environment. When both parties demonstrate care and concern for each other's needs, employee engagement, productivity, and satisfaction are greatly increased (Mani & Nadu, 2011).

Healthy ergonomics creates a win-win situation for everyone!

REFERENCES

Lerner, D., & Henke, R. M. (2008). What does research tell us about depression, job performance, and work productivity? *Journal of Occupational and Environmental Medicine*, 50(4):401–10. DOI: 10.1097/JOM.0b013e31816bae50.

Mani, V., & Nadu, T. (2011). Analysis of employee engagement and its predictors. *International Journal of Human Resource Studies*, 1(2):15–26. DOI: 10.5296/ijhrs.v1i2.955.

WHO. (2000). *Mental Health and Work: Impact, issues and good practices*, World Health Organisatation, Geneva. https://apps.who.int/iris/bitstream/handle/10665/42346/WHO_MSD_MPS_00.2.pdf

Work Health and Safety Act 2012. (South Australia). V.3.10.2019 (Austl.). Retrieved from https://www.legislation.sa.gov.au/__legislation/lz/c/a/work%20health%20and%20safety%20act%202012/current/2012.40.auth.pdf

3 Return-to-Work and 24/7 Warehouse Operations

Wenqi Han

CONTENTS

ERGONOMIC INTERVENTION FOR AN OFFICE WORKER WITH SPINAL CORD INJURY

Disability includes several aspects, such as cognitive, developmental, intellectual, mental, physical, and sensory deficits, and any mixture of these that affect function. Disabilities affect personal lifestyles and can be genetic or befall a person during their life journey. In Singapore, people with sensory and physical disabilities constitute about half of the disability group, with the remainder including intellectual disabilities and autistic spectrum disorders.[1] Also, worker disabilities can be classified to differentiate and facilitate employment matching with their job specifications. This is done when assessing the medical condition through sensory examination, motor examination, and determination of single neurological level, complete or partial spinal cord injury (SCI), and grading of impairment.[2]

This case is about my experience in conducting an ergonomics study and developing a return-to-work programme for a previously fit male weighing 75 kg and 170 cm stature with no underlying medical conditions. Due to privacy and confidentiality reasons, he will be referred to under the alias "James". James was a 30-year-old construction planning engineer. In 2017, he was involved in a hit-and-run collision while crossing the road during his commute to work one morning. James suffered a spinal shock and was diagnosed with Asia Impairment Scale (ASIA), Grade C, T12 of the thoracic spinal segment, a SCI. He required wheelchair assistance and was admitted to a rehabilitation unit in the hospital. After two weeks of hospitalisation, James underwent weekly physiotherapy.

As a Workplace Safety and Health (WSH) consultant, I was approached by the Human Resource Manager and Operations Director to attend an informal meeting. They wanted me to perform an ergonomic study and suggest measures about the return-to-work programme of our injured colleague. Since James was on a 90-day

---

[1] https://www.msf.gov.sg/media-room/Pages/Total-number-of-persons-with-disabilities-in-Singapore.aspx
[2] https://www.spinalinjury101.org/details/levels-of-injury

DOI: 10.1201/9781003349976-3

medical leave, this was the implied deadline to craft a plan to implement workplace adjustments. People with SCI face significant environmental barriers that affect their social participation and employment (Tsai et al., 2017). In general, an ergonomic assessment should be conducted to assess the worker rehabilitation needs by considering the medical diagnosis, task requirements, and workstation layout.

I considered the medical diagnosis of James to gain a basic understanding of his needs. People with T1–T12 paraplegia have nerve sensation and function of all their upper extremities. They can transfer independently and manage bladder and bowel function. Individuals with a T10–T12 injury have better torso control than those with a T2–T9 injury, and they may be able to walk short distances with the aid of a walker or crutches.2

During my research and reading, I considered two options taught during my university studies: the ASIA International Standards for Neurological Classification of Spinal Cord Injury checklist,3 where the sensory and motor functions are scored, and Spinal Cord Injury Research Evidence (SCIRE) Outcome Measures Toolkit.4 The utilisation of both tools was agreed upon after consultation with the company doctor, who had also assisted me in conducting a clinical neurological examination on James to rate his sensation and muscle function during his medical leave. The doctor informed me that James would still need the assistance of a wheelchair after his 90-day medical leave and return-to-work.

I conducted an ergonomic investigation on James's desk to evaluate compliance with the Singapore Standard SS514:2016 Code of Practice for Office Ergonomics (Singapore Standards Council, 2016). This is a mandatory requirement that governs office ergonomics in the Singapore context. This assessment was essential to discover opportunities to redesign the workplace. As part of the investigation, I took pictures and measured the existing desk dimensions (L: 1,200 mm × W: 600 mm × H: 750 mm) and the room's layout (L: 6,100 mm × W: 3,600 mm). Also, I called James on the phone to understand his substantive and routine work activities before deciding whether a workplace redesign was needed. His regular work activity included using a computer for typing reports (6 hours/day) and walkabouts to collaborate with others or obtain updates and progress details (2–3 hours/day).

I compiled the data and information needed to inform my recommendations: wheelchair dimensions and functions based on online research and information from the company doctor, and a review of James's previous work area, per the Accessibility Code 2019 by the Building Construction Authority of Singapore (Building Construction Authority, 2019). After contemplating the human, logistical, and infrastructure parameters, I synthesised the information and prepared a proposal for managerial approval. The proposal included a to-do list and a breakdown of the critical considerations for redesigning the workspace to accommodate James. During the co-design process, I also consulted with James. His only concern was the movement and manoeuvring of his wheelchair within the workplace.

---

3 https://www.physio-pedia.com/American_Spinal_Cord_Injury_Association_(ASIA)_Impairment_Scale
4 https://www.physio-pedia.com/Spinal_Cord_Injury_Outcome_Measures_Overview

TABLE 3.1

Principal Changes in James's Work Design

Previous Work Activity	Revised Work Activity	Duration Changes	Main Workplace Changes
Use of computer for typing reports	Unchanged	Reduced from 6 to 5 hours	Redesign of workspace and desk to help James fit into the environment
Walkabouts to collaborate, obtain updates and progress details, etc.	Use of computer system to monitor worksite progress through artificial intelligence	Unchanged (2–3 hours)	Use CCTV instead of site walkabouts to monitor progress and receive updates

The proposal was compiled based on no changes in job functions and scope of work, meaning a return to his substantive duties in full, and resulted in the equipment-related work design changes (Table 3.1).

Pertinent to the workplace redesign, I considered the (1) stationary dimensions of an adult wheelchair, (2) manoeuvring dimensions of an adult wheelchair, (3) dimensions of the office layout, (4) dimensions of the common access corridor, (5) dimensions of adult wheel chair side reach, (6) dimensions of adult wheelchair front reach, and (7) Smart Commodity CCTV (mounted on tower crane mast and every storey) for the identification and real-time monitoring of work area (360°). The final proposal included (1) power socket outlet – height at 900 mm, (2) wall-mounted adjustable monitor screen at 1,200 mm (360°), (3) two-way lighting switch – height at 900 mm, (3) air-conditioning remote control – height at 900 mm, (4) emergency press button – height at 900 mm (linked to the reception desk with sound alarm and SMS alert), and (5) wall-mounted adjustable work desk of height: 850 mm, depth: 600 mm, and length: 1,200 mm.

The management approved the proposal. As the changes and investments required for the redesign were not major, the management started the procurement process within two days while enquiring about possible government grants to defray the costs. The minor renovation took two weeks to complete with a cost of $13,500, which, in addition to the design features presented above, included:

- Installation of a wardrobe (height: 900 mm, width: 300 mm, and length: 3,600 mm),
- Alteration to partition wall to install conceal conduits for lighting switch and power socket outlets,
- Electrical wiring works,
- Data and CCTV cabling works, and
- Installation of an adjustable bracket for the desktop screen.

Interestingly, managers felt that they should invest in technology to aid everyone, not only James, in the era of technological advancements. They realised that installing the CCTV to view updates instead of performing site walkabouts would be a risk reduction method to minimise exposure to various hazards in physical worksites for other employees. Physical walkthroughs can hide risks like slips, trips, and falls (STPs), which could be eliminated by minimising the exposure to the hazards by deploying an intelligent CCTV for meetings and discussions through real-time connections to the worksite. This could positively affect safety, productivity, and efficiency, and the adoption of smart innovations and technologies to ease task load could be perceived as beneficial by the workforce. Of course, someone must still balance remote work with on-site physical presence to engage with employees, build trust, and empower a psychologically safe work environment. Equally important, the opportunities to attend worksites and not just monitor them from a distance should be given to workers to avoid long sedentary activities and promote physical fitness.

The management proceeded with this initiative, but the CCTV vendor reminded us about the option for personnel to work from home and access an online platform to co-utilise the CCTV with Internet access and password upon management approval. An option was offered for an add-on to the bundle on Apple's Siri, a virtual assistant in Apple Inc.'s operating systems. Nonetheless, those two features were not installed due to considerations about the feelings of James. We intended for workplace inclusion and protecting the dignity of the recovering colleague. Working from home could be promoted when the information technology, Internet connection, and organisation's system accessibility are enabled. This could encourage flexible work too.

I felt a sense of achievement after the renovation was completed. I reflected on the experiences and challenges that this project presented: the initial meeting with the Human Resource Manager and Operations Director, the concerns about the worker, the need for an objective and scientific-based approach, the conduction of the ergonomic investigation, and the completion of the workplace redesign. The insights that I gained were valuable. I did not have prior experience in handling return-to-work programmes and performing ergonomic investigations of workspaces. Thus, it was challenging to gather all necessary information in a limited time and prepare the proposal.

I had thoughts of inviting an industrial ergonomist to provide professional advice, but time constraints did not allow for that. Also, this could be partly attributed to my own exploratory personality with perseverance to confront new challenges and a "willing to do" attitude to complete the proposal. After all, I was employed based on my basic knowledge of ergonomics as part of my bachelor's degree in safety, health, and environmental management. I was trained and assessed in this course to demonstrate competency in carrying out ergonomic assessments by using the tools I applied to this case. Furthermore, it is not a stipulated requirement to engage an industrial ergonomist for such evaluation and planning. Nevertheless, time and financial constraints are areas of concern, meaning that the involvement of an industrial ergonomist may not always be feasible. Indeed, it would be worth examining when and how to mandate the engagement of ergonomists in specific cases and possibly recruit industrial ergonomists to contribute to WHS and well-being aspects.

Return-to-work is a humane approach, an obligation, and a duty of care for employees. Workplace adjustments are individualised solutions that enable persons with disabilities (PWD) to carry out tasks and remain productive. Adapting the work environment, providing assistive devices, modifying working schedules, and redistributing non-essential tasks to other workers are some examples of reasonable accommodations. Implementing those is a vital path for increasing employment for PWD who often encounter difficulties with functional movement or exertion demands related to routine work.

In general, a thorough assessment supported by management can realise work redesign that considers PWD and provides an inclusive workplace that offers more agility for others once the learnings are learned. In this case, the implementation of CCTV technology led to efficiencies and risk reduction for all team members involved. It is conceivable that workplace adjustments to cater for the needs of PWD during their return-to-work could challenge employers if they are ill-prepared. Unfortunately, some organisations might underestimate the value of return-to-work schemes and miss the opportunity for workplace inclusion of injured employees. Maybe, in such organisations, there is a perception that injuries are temporary, and healing shall happen only at home or elsewhere (e.g. sheltered workshops).

However, a more proactive approach is warranted as PWD represent a significant percentage of the population and can be imaginative people with skill sets that should be recruited and retained in a workforce. For example, building designs shall incorporate accessibility to the workplace and other facilities (e.g. car parks, ramps, lifts, and washrooms). Furthermore, apart from PWD, with a retirement age stipulated at 67, an increasing worker population group clearly requires ergonomic considerations for an accessible workplace. With a fast-ageing population, we need interventions initiated by any level of stakeholder and a coordinated response among government authorities, company decision-makers, renovation contractors, Human Resource managers, industrial ergonomists, and WHS professionals.

In Singapore, for individuals over 50 years old, around 13.3% are considered disabled; between ages 18 and 49, about 3.4% are disabled; and 2.1% of children under 18 are disabled in Singapore. Of those with a disability, around half are considered physical or sensory disabilities. One in 68 children in Singapore has been diagnosed with autism, and this number has increased over time. About 5 to 6% of children born in Singapore have developmental problems of various types, and 0.55% of the Singapore workforce has a disability of some kind.1

From a self-reflection viewpoint on this ergonomic intervention project for an office worker with SCI, I would like to share my thoughts and critical insights gained:

- I am someone with high expectations, and I feel there is an immediate need to pursue further education, focusing on human anthropometry, human–machine interface, and ergonomics of tool design to enhance my knowledge to equip myself for more uphill challenges.
- On top of the physical ergonomics, we also need to have the skill sets to understand cognitive ergonomics, mental health, and well-being of persons returning to work.
- The psychosocial risks arising from constraints imposed by organisational safety climate and culture, perceptions of individuals, and diverse human factors characteristics can significantly differ across space and time.

Based on the above, I am currently exploring a career development plan through continuing education. Another option I consider is self-study of related research and books to enrich my knowledge base. I believe that honing my skill sets is similar to sharpening the saw. This is also one of the habits of successful people indicated by Steven Covey in his book – *7 Habits of Highly Effective People* (Covey, 2009).

ERGONOMIC ASSESSMENT FOR 24/7 WAREHOUSE OPERATIONS

In this case, I worked as a WHS manager in a logistics and warehousing company. The past injuries and incidents in the logistics and warehousing industry supported the need to implement effective plans to mitigate risks and improve the well-being of employees. Pickers are exposed to the risk of collision with forklift trucks. Accidents involving forklifts and trucks are rising in the logistic and warehousing environment (Lam, 2017; Min, 2017).

This organisation had a headcount of 1,795 workers, 1,200 of which were drivers, 55 workshop mechanics, 150 administrative staff, 25 management personnel, 15 salespersons, and 350 warehouse staff. On my first day at work, after my orientation and introduction to the organisation, I browsed through the files at my work desk left by my predecessor. I found documents about risk assessment and safe work procedures for forklift operations. I asked my colleagues sitting nearby, two with management roles and one warehouse staff, about their perception of WHS at the company. Essentially, they replied, "just do the bare minimum; if there is no need to change, the better". I also noted that there was no established WHS management system, like policies, procedures, training materials, and reporting methods.

On my second day at work, I asked the warehouse manager about WHS in his area of responsibility. He explained that the authorities only required a risk assessment and safe work procedure, audited once every three years. Other than that, the manager stated, they did not need to do or have more, and everyone knew that safety was paramount. When I asked about the incident rates, he replied, "touchwood, we had eight minor injuries last week, which we consider normal". I informed the warehouse manager that I would propose a WHS action plan for him and the management to consider. The meeting ended with him advising me, "Don't do too much to hinder the work progress, and don't stop any work unnecessarily".

Two days later, at about 09:15 am, my phone rang. It was the warehouse manager. "Morning Han", he said, "are you available to attend a short meeting at the 2nd storey meeting room at 10:00 am regarding the recent direction by the management on the 24/7 warehouse operations?". After the call ended, I tried to brainstorm what they might need from me. I thought that I should have some information handy about the risk profile of the workforce based on their job types. I scribbled on my notebook the immediate thoughts that emerged from my academic studies on ergonomics and related, especially, to sedentary and non-sedentary job activities.

Additional thoughts included information about the employees' lifestyle (e.g. eating habits, weight management, and tobacco use), which I jotted down. Sleep was the next concern because I wondered whether prolonged sleep deprivation or sleep pattern changes and their possible impacts could be an issue for the workforce. This preparation took me about 40 minutes, including data collection and searching for an appropriate survey tool. I found that operations in the warehouse typically ended at about 22:00 each day, after significant overtime. Presumably, this could have triggered management to consider a 12-hour shift and 24/7 operations. I also found the Basic Health Survey Checklist developed jointly by the Workplace Safety and Health Council and Ministry of Manpower of Singapore (WHSC, 2018).

I quickly proceeded from my work desk to the meeting room. I did not want to be late. The meeting started with topics about workforce, work pass issues, accommodation, food, insurance, payroll, utility bills, upcoming contract for orders, need to ramp up outgoing cargos of extra 50%, and expected revenue of 25%. I waited for my turn, which I saw on the projector screen with "Health & Safety for 24/7 Warehouse Operations" with no content. I asked the warehouse manager, "Can you clarify what is expected of me as a WHS professional to contribute to this topic?" There was complete silence in the meeting room for a few minutes while colleagues looked at each other.

Following those awkward silent moments, I shared all the key concerns from a risk management point of view based on my earlier brainstorming and cursory research. I explained that I could design a checklist to gather data from individuals with assistance from the Human Resource Manager, company doctor, and line supervisors. Then, I would compile and submit an Ergonomics Risk Management report. The points and recommendations I shared were recorded in the meeting minutes.

After lunch that day, I returned to my work desk, turned on the power of my desktop, and started typing a draft proposal with the key points that I had written in my notebook earlier in the morning. To gather data and understand the profile of the workforce, I built a survey form with questions about demographics and individual characteristics (e.g. age, gender, weight, height, (non)sedentary job tasks, eating, sleeping and smoking habits, mental health, and recent/current medical conditions), effects of the work system (e.g. fatigue, work pace, postures, stress), awareness about accidents and WSH, and environmental factors (e.g. temperature, humidity, lighting, ventilation, noise, vibrations). The Basic Health Survey Checklist mentioned above was rather generic. Therefore, I thought of enriching and adjusting it to accommodate more contextualised items applicable to the warehouse environment. There was no time, and nobody with the necessary knowledge was immediately available to review the draft checklist.

The survey form was emailed to the Human Resource Manager, and it was printed and distributed to everyone through the line supervisors and heads of departments. The entire workforce participated in the survey at a designated workstation with different time slots and a short briefing on the instructions and rationale of this survey. The process was also assisted by three interpreters of the native languages, namely Malay, Mandarin, and Tamil. As Human Resource is being perceived as a department with authority and representing the directions of top management, the 100% response rate was somewhat expected.

The responses took about four days to come back to me, and data compilation took another two days. Nonetheless, I felt that the psychological safety within the workplace to express and accept factual feedback was at a low level. Therefore, the reliability of the data collected could have been compromised due to fears of a backlash. Hence, I carried out a walkthrough at the warehouse to evaluate the environmental factors for verification purposes. The 20-year-old facility was run-down and not well-maintained, and humidity and temperature in the warehouse were a concern. Most of the lighting was coated by dust. Ventilation was poor, and the environment was also dusty due to the poor housekeeping and regular traffic of vehicles. The engine noise from transportation vehicles was also a factor, as prolonged exposure could cause hearing problems.

The information was recorded and used in my risk assessment. As a WHS practitioner, in the light of these observations, I asked myself, "how did such an organisation exist? And for such a long time?" I felt that improving the work environment could be a new challenge that I would like to conquer by gaining buy-in from the decision-makers and stakeholders. There was another thought which came across my mind. During a job interview, how about everyone seek permission from the interviewers to have a short tour around the operations area to catch a glimpse of what kind of workplace they are going to work in?

The survey results revealed that 60% of the workforce were locals, aged above 45 years old, and 40% were foreigners, between 20 and 30 years old. Below, I list the main findings from the survey and some data I distilled from annual medical check-ups performed by the company doctor. I have made indicative notes about their importance:

- 18% of the employees performed sedentary work without regular breaks for physical activities. Those workers could be exposed to a higher possibility of heart diseases, diabetes, obesity, high blood pressure, stroke, depression, and chronic cancer due to physical inactivity.[5]
- 75% of the workforce fell under the obese category, and 70% of the workforce had high blood pressure and did not often exercise due to their long working hours and overtime. Obesity combined with environmental stressors leads to several chronic illnesses. Poor eating habits may range from irregular hours of eating to disproportionate amounts of food intake, including habits like eating dinner late.[5]
- The travelling time between work and home through shift work took more than 65% of workers' day, on average 16 hours. Notably, the warehouse was very far from the nearest public eatery facility, which is 2 km away. Shift work deprived employees of exercising options as they needed additional rest after the long working hours. Several gastrointestinal (GI) disorders such as ulcers are more common across shift workers than other employee groups. Also, shift workers are more likely to suffer abdominal pain, gas, diarrhoea, constipation, and nausea caused by a decrease in appetite and indigestion.[6]
- 85% of the employees were smokers. Tobacco consumption leads to heart diseases, high blood pressure, stroke, and other chronic cancers. Smokers suffer from respiratory diseases, severe airway damage, emphysema, and chronic bronchitis.[5]

---

[5] https://medlineplus.gov/healthrisksofaninactivelifestyle.html
[6] https://www.medicalnewstoday.com/articles/list-of-digestive-disorders#types

- 75% of the workforce felt overloaded with tasks and stressed by management's high demands and expectations. Although small amounts of stress can be reasonable and motivate performance, chronic stress can lead to serious health problems. Multiple daily challenges (e.g. fast-paced environment, meeting deadlines) together with other internal and external environmental stressors (e.g. noisy work environment, poor workplace lighting, family commitments, personal matters) can lead to excessive and prolonged stress.
- 85% of the workforce complained about inadequate rest as they are required to work overtime till 10:00 pm on most of the days. As restoration and renewal of the human body occur during sleep, victims of sleep deprivation often suffer from physical and emotional disturbances (Medic et al., 2017). Also, a disruption of circadian rhythms, which regulate the "normal" awake and sleep cycles, can make people sleepy, somnolent, or unable to sleep when possible.[5]

I started drafting my proposal to suggest how to eliminate or reduce for all employees the WHS risks and promote health through workplace interventions. I proposed solutions by consulting the company doctor, based on my previous experiences and ideas from the literature (Table 3.2). As there were time constraints and this was the first WHS assessment in this organisation, I did not consult with the workers.

I submitted my proposal, and the managers jointly decided to proceed with the 24/7 operations of the warehouse. However, they also decided to shelve my proposal due to their perception of high implementation costs. They reiterated that having a risk assessment and safe work procedure was good enough. Implementing the recommended risk control measures was not perceived as necessary because they would not be required to demonstrate compliance with the authority at their next visit.

My reflection on this failure to gain the buy-in of my WHS improvement proposal generated mixed feelings. I wondered, who should be responsible for WHS? Why did they reduce WHS to just documentation of a risk assessment and safe work procedures? What was the role of the auditors? Had they performed physical inspections? Had they interviewed any workers? How could the company have passed the audit with so many issues? Through my reflection, I realised we should assess an organisation's readiness for change by understanding management commitment, resource and support allocated, and existing barriers that workers face. Last but not least, the perceptions of all stakeholders that affect the organisation's psychosocial safety climate and culture will drive the intent and commitment to invest in the safety, health, and well-being of the workforce.

When no mishap occurs, the definition of "reasonably practicable" measures is not questioned. Promoting greater ownership to encourage voluntary efforts is good for proactive safety but can be perceived as unproductive from an outcome-based or purely productivity perspective. Many other questions emerged! Where is the leadership, and what is their commitment to ensuring WHS? Will they be ever ready to integrate WHS into work processes? Why did the previous WHS incumbent leave? What should I have done differently to gain buy-in from management on my proposal? How would they handle civil litigation claims if any employee suffers illness

TABLE 3.2
Workplace Risks and Strategies

Risks	Strategies (and Indicative Sources Where Applicable)
Poor health due to sedentary job tasks.	Alarm reminder every 2 hours to remind administration office staff to do stretching Workspace redesign for both healthy staff and people with disabilities Internal gym facilities Internal clinic facilities Quarterly body check-up regime Consider "Keeping fit"/Align Body Mass Index (BMI) as one of the key performance indicators (KPIs) Assessment to understand current office work conditions and create a sustainable work environment for administration staff. Some of the suggested assessment tools were the Rapid Upper Limb Assessment (RULA), Rapid Entire Body Assessment (REBA), and American Spinal Injury Association (ASIA).
Health impacts due to poor eating habits and weight management	Wearable gadgets to monitor the health status Staff cafeteria with "healthy" meals provided for all staff free of charge Employment of dietician and nutritionist to create tailor-made menus together with chef Displaying posters to encourage eating healthy and keeping fit Regulate duration for meals (e.g. lunch 11:30–13:30 and dinner 17:30–19:30) Employees to take their meals before every work shift ends to ensure they do not eat late after they go home
Poor health due to tobacco consumption	Smoking restrictions – restrict smoking corner in company premises Smoking cessation clinic, hotline, and counselling Nicotine replacement therapy (NRT) and antidepressants such as bupropion
Effects of stress	Surveillance regime on medical leave rates Quarterly cohesion events to knit bonding of colleagues, management, and present appreciation awards Annual holiday retreat trips Medical benefit scheme for family members of employees Counselling sessions
Disorders due to inadequate sleep	Company transport to and from the doorstep Company laundry services Resting area for employees during short breaks Annual sleep test regime Use wearable gadgets to monitor employees' blood pressure, heartbeats, etc. Sleep intervention programme
Effects of adverse environmental factors	Regular luminaires level monitoring and maintenance of lighting Designated loading area with mechanical ventilation systems Regular air particle monitoring and maintenance of ventilation systems Vehicles to switch off their engine if idle for more than two minutes. Noise monitoring Hearing Conservation Programme and ensure all personnel are equipped with hearing protectors Hearing tests for all

due to a poor work environment and prolonged work hours? What are the business continuity strategies and plans if the authorities uncover those unsafe working conditions and breaches of employment law requirements?

My conclusion of this case is that a WHS management system or plan should be embraced, supported, and effectively implemented. We are paid as an employee to advise management on aspects of WHS legal compliance and control of all foreseeable risks. However, the decision still lies in the managements' hands. To the best of our knowledge, we do what we can to recommend WHS initiatives and programmes for implementation. However, the organisation's cultural maturity journey might be much slower than our expectations. Business continuity management and WHS are critical elements to remain competitive in the market, but the latter might not always be a priority. Leadership and commitment, training to build competencies, and resources necessary to run a 24/7 warehouse operation within a safe and healthy working environment require investments that managers might not see necessary if not enforced.

Looking back, I pondered if I were to restart again on this assessment for the 24/7 operations of the warehouse, how could I do it differently? I may do it in a more paced fashion by obtaining and processing currently available data from the heads of departments instead of administering a new survey. An organisation without data from past years to inform a trend analysis indicates few possibilities. First, there might not have been much attention to WHS due to a lack of serious incidents. Second, there could be poor competencies of internal WHS staff. Third, management might be only interested in profit- and delivery-based outcomes. Thus, instead of drafting a full-scale proposal, I could have suggested an improvement plan with immediate actions to make the existing environment safer before setting sails for a bigger mission.

REFERENCES

Building Construction Authority. (2019). *Code on Accessibility in the Built Environment 2019*. Singapore: Building Construction Authority. Retrieved from https://www.corenet. gov.sg/media/2268627/accessibility-code-2019.pdf

Covey, S. R. (2009). *The 7 Habits of Highly Effective People*. New York: Rosetta Books LLC. Retrieved from https://fb2bookfree.com/uploads/files/2020-05/1590810411_the-7-habits-of-highly-effective-people.pdf

Lam, L. (2017). *Logistics Firm Fined $80,000 Over Accident Where Forklift ran Over Worker's Legs*. Singapore: The Straits Times. Retrieved from https://www.straits times.com/singapore/logistics-firm-fined-80000-for-forklift-accident-that-ran-over-workers-legs

Medic, G., Wille, M., & Hemels, M. E. (2017). Short- and long-term health consequences of sleep disruption. *Nature and Science of Sleep*, 9, 151–161. DOI: 10.2147/NSS.S134864.

Min, C. H. (2017). *Lorry Driver Crushed After Forklift Drops Steel Bars in Fatal Industrial Accident*. Singapore: The Straits Times. Retrieved from https://www.straitstimes. com/singapore/lorry-driver-crushed-after-forklift-drops-steel-bars-in-fatal-industrial-accident

Singapore Standards Council. (2016). *SS 514:2016 Code of Practice for Office Ergonomics*. Singapore: Singapore Standard Council. Retrieved from https://www.singaporestan-dardseshop.sg/Product/SSPdtDetail/b558a5ee-9bc3-4b0d-a96b-6e518a990f21

Tsai, I.-H., Graves, D. E., Chan, W., Darkoh, C., Lee, M.-S., & Pompeii, L. A. (2017). Environmental barriers and social participation in individuals with spinal cord injury. *Rehabilitation Psychology*, 62(1), 36–44. DOI: 10.1037/rep0000117.

WHSC. (2018). *Basic Health Survey*. Singapore: Workplace Safety and Health Council. Retrieved from https://www.tal.sg/wshc/-/media/TAL/Wshc/Programmes/Files/BHS_v3_22072019.pdf

4 Designing a Visually Comfortable Workplace

Jennifer Long
Certified Professional Ergonomist, Australia

CONTENTS

As a teenager, I spent many hours contemplating what my future career would be and how I could change the world or, at least, make a difference for some of the people who lived in it. I was certain I wanted a science-based career but was torn between engineering and optometry. I originally placed engineering on the top of my university admissions form, but at the 11th hour swapped it to optometry because I felt a greater calling to help people to see better. It was such an angst-ridden time making a binary choice for my career path when I was 17 years old. Who would have guessed that a career path commencing in optometry would eventually intersect with the engineering/built environment world? But it did. And it was through this intersection of careers that I found an avenue for making a difference in this world.

In the early 1990s, there was a vast increase in the number of people using a computer at work. As an optometrist working in clinical practice, I observed a corresponding increase in the number of patients who presented for an eye examination reporting sore eyes and headaches which they attributed to using a computer. But when I examined their eyes, there was often nothing wrong. I was intrigued, "What's going on with these patients?" I started to question these patients about their work and found that the most likely problem was the way that their computer was set up, the display of their information on their monitor, the quantity and quality of office lighting, or the fact that they did not take rest breaks during the day.

Unfortunately, visual comfort and ability are often overlooked components in workplace and equipment design. It's not until people occupy the space or start using

DOI: 10.1201/9781003349976-4

equipment that problems are discovered, such as glare causing headaches, computer displays positioned in locations that promote awkward postures, or computer interfaces with character elements that are difficult to see. Thus, frustrated with my inability to help these people, I decided that I might be more effective if I could address these problems at their source, that is, in the workplace.

To achieve my new aim, I went back to university and studied to become an ergonomist. I anticipated that instead of giving advice to individuals in the optometry consultation room, my new career would enable me to give advice to individuals in the workplace and, ultimately, solve their visual discomfort. My second career as a consultant specialising in visual ergonomics commenced in 2006. My work broadly fell into two categories:

- Providing advice to help people who were experiencing visual discomfort due to the design of computer interfaces.
- Providing advice to help people who were experiencing visual discomfort due to the work environment.

By about 2008, I realised that I wasn't making the difference I had envisaged. The designs and equipment used by people in the workplace had already been built, purchased, or installed. My advice was often limited to suggesting retrofit solutions which were potentially expensive and not necessarily a good solution. I started to wonder, "Wouldn't it be more effective to identify and solve problems BEFORE designs are built or workers occupy the space?"

The good news is that I have since had the opportunity to contribute evidence-based visual ergonomics advice to the design process. It hasn't all been plain sailing. In this chapter, I present two examples of my work. The first case example describes my less-than-successful attempts to influence the visual design of computer interfaces. The second case example describes my more successful attempts to influence visual ergonomics design elements within control rooms. Had I been asked to explain the difference between the successful and less-than-successful projects, I would nominate "communication" and "egos".

VISUAL DESIGN OF COMPUTER INTERFACES

I was contacted by the health and safety (H&S) manager of an office-based organisation where employees performed data entry tasks for most of their working day. Many employees within the organisation were upset with a new computer program that had been implemented across the business. There was disquiet because the font size on the display was less than 2 mm high, and some employees found themselves leaning forward across their desk to read from the display. Another source of irritation was the fact that the display colours were visually uncomfortable and could not be modified (e.g. black font on a dark green background). Consequently, many employees reported eye strain and headaches by the end of the working day.

"Can the workers enlarge the font size, so it is easier to read?" I asked.

"No", the manager replied, "The interface elements have a fixed design".

"Can you change the computer monitors or the screen resolution?"

"No", the information technology (IT) manager informed me, "The program has been designed for use on a specific size monitor and screen resolution. If we change the monitor or the screen resolution, it only makes the problem worse".

"Can you describe the problem to the interface designer and ask them to modify the program?"

The IT manager was adamant. "No, we have already purchased the program. We cannot make major design changes like that".

"Can you revert to using the old program while you sort out a solution?"

Both the H&S manager and IT manager were aghast. "Are you kidding??? The business has just spent several million dollars on this new program! It has to work!!!"

I wish that the above were an isolated anecdote. Unfortunately, variations on these conversations litter my visual ergonomics consultancy career. Nevertheless, I thought I struck gold when I was invited by the site ergonomist of an industrial plant to speak to employee engineers about the computer displays used on site. These engineers had some influence over the computer interface design. The site ergonomist thought it would be useful if I could explain the rationale behind the interface design elements so that the engineers were more informed when making design decisions.

During the education session, I shared visual ergonomics pearls, such as why the font needs to be a minimum size so employees don't need to lean across their desk to read from the display (Rempel et al., 2007). Also, why it's not a good idea to use red font on a blue background[1] (Travis, 1991). At the end of the session, one of the engineers exclaimed, "I wish someone told me this years ago! Now the design rules make more sense".

However, my connection with the industrial site ceased when the site ergonomist changed her employment. As such, I could not know what changes the engineers were able to implement with their computer interfaces. Nevertheless, I have had similar interactions with engineers in subsequent consultancy projects across a range of industries. When I've explained the rationale for the design rules, their responses have been almost unanimous, "I wish someone told me this years ago!"

Interestingly, although organisations may employ engineers and IT specialists as permanent staff to help design and implement computer interfaces within their business, I've observed that the design influence of these employees is often limited. For example, while working on a project to redesign a control room, there was an engineer who was assigned to work with the software company who were developing the computer interface. The engineer had identified potential problems with the interface that was going to affect its usability and the visual comfort of the control room operators. The problems were unresolved, despite several conversations and meetings with the software developer.

The engineer thought he might have more success resolving the potential interface problems if he armed himself with some solid facts. He attended the next meeting with the software developer equipped with Hollifield's High Performance HMI Handbook (Hollifield et al., 2008) and informed with the visual ergonomics knowledge learnt from me. Unfortunately, these resources were insufficient to penetrate the

---

[1] This sets up chromostereopsis, an optical effect where the page appears to shimmer, caused by the fact that our eyes are unable to simultaneously focus on the red and blue colours.

veritable steel wall surrounding the software company and its representative, who repeatedly uttered, "This is the base design. If you want to change the base design, then it will take several years for a dedicated team to do this". Of course, it would also require vast amounts of money to create a bespoke solution. The consistent message implied by the software company was that other clients weren't complaining about the base product, so why should they modify it? I wish this was another isolated anecdote, but variations on this scenario have also been common throughout my visual ergonomics consultancy career.

Maybe it is naïve on my part, but it mystifies me why software development companies are seemingly unaware of basic visual ergonomics knowledge that can improve the usability and comfort of their designs. This knowledge is in the public domain and published in standards, books, and even on the Internet. I've asked software developers while working on projects, as well as people I know socially who work in the industry, "What are the barriers to implementing good design elements?" My question has usually been met with a bemused look and a shrug of the shoulders to convey "I don't know".

My hope is that the small amount of work that I have done with engineers within companies, together with the recommendations I've written in reports, will be adopted in future iterations of software used within those organisations. Maybe, little by little, this will improve the visual comfort and ability of the people who use these systems. Ideally, I would like to see an industry-wide approach to improving the visual design of interfaces, for example, by including visual ergonomics education within engineering curricula so that software engineers understand the rationale for design decisions. This could be coupled with greater consumer awareness of good visual design elements, for instance, by including visual ergonomics criteria in the design and procurement phases of products.

VISUAL ERGONOMICS DESIGN ELEMENTS OF CONTROL ROOMS

FIRST ENCOUNTERS

My first encounter with a control room was early in my consultancy career. I was contacted by the H&S manager of an industrial plant because operators in the control room were unhappy. Some operators described difficulty seeing the displays and visual discomfort by the end of their 12-hour shift. I was engaged to give advice on how to improve the visual comfort and ability of the control room operators. I was expecting an ugly environment, but, to my surprise, the control room was well-appointed and aesthetically pleasing. In hindsight, this was one of the better control rooms I have visited during my career. On the other hand, unfortunately, the problems described by the control room operators were not easy fixes.

The operators were using two-tier, vertically stacked displays on their console. The upper displays were above eye height, necessitating a head-tipped-back posture to read from the displays. This is not an ideal physical arrangement because it can contribute to neck and shoulder discomfort. It is also contrary to ergonomics best practice, which recommends that frequently used displays should be below eye height (Burgess-Limerick et al., 2000; ISO, 1999; Villanueva et al., 1996). The uncomfortable posture was even more pronounced for operators aged over 40 years who were wearing general-purpose progressive lens spectacles or general-purpose bifocal spectacles to correct the optical effects of presbyopia.

Presbyopia is a normal age-related change that reduces the eye's ability to easily adjust focus between close and far distances. People typically notice difficulty seeing objects up close, and it is easily remedied with reading spectacles. General-purpose progressive and bifocal spectacles are ideal for people with presbyopia who want one pair of spectacles that allow them to see objects at a close distance (e.g. for reading) and objects at a far distance (e.g. for driving). The near vision zone is in the lower portion of the spectacle lens, enabling the wearer to simply look downwards to view hand-held objects such as books or smartphones. However, the location of the reading zone in the lower portion of the lens promotes a head-tipped-back/chin thrust forward posture when wearers try to read from a desktop computer display (Martin & Dain, 1988). Consequently, general-purpose progressive and bifocal lenses are contraindicated for desktop computer tasks (Long, 2019).

The optimal solution for the control room was to redesign the computer interfaces and replace some of the displays on the consoles. This would ensure a more comfortable height for operators seated at the console but was beyond the budget of the business at that time point. I discussed possible solutions with the manager, including repositioning frequently used display content onto lower tier displays that were below eye height. We agreed my report should provide advice for what the business should consider next time they built a control room, as well as advice on how to manage the existing problems.

My report also included recommendations for alternative types of spectacle lenses that operators should wear while working at their console to reduce their risk of neck and shoulder discomfort from viewing displays above eye height (Long, 2019). While I know that my report was focused heavily on optometry-type advice, I felt that I had made a step forward in solving vision problems at their source. However, in making this step forward, I also discovered that "the source" was not really the workplace. It was higher up, that is, with the people who designed the control room. My lingering question was "How can I work with someone to design better control rooms?"

Wishes Can Come True

For several years, I continued to provide visual ergonomics advice in control rooms that were already built and operational. Each time, I provided advice on what I considered the business should incorporate when they upgraded their control room as well as simple retrofit solutions that would improve the comfort of operators working in the room. One day, I received a phone call from a control room architect. He had read a report I had submitted to a mutual client. He shared my philosophy that it is better to design out problems before they are built into a system. I was invited to work with him and another ergonomist to provide very early schematic design advice for control rooms.

My role in the team was threefold: provide visual ergonomics evidence-based information that would inform good design, help discover the stakeholder requirements for the control room, and assist facilitating participatory ergonomics workshops with the control room stakeholders, including control room operators, managers, ancillary staff, IT support staff, and facility managers (Long et al., 2019). Participatory ergonomics is an interactive process whereby stakeholders are provided with evidence-based ergonomics knowledge and design principles relevant to their workplace. Through facilitated workshops, the stakeholders are supported and encouraged to apply their own knowledge of the work and work tasks to solve workplace ergonomics problems (Burgess-Limerick, 2018). The education component of this process was critical because it helped the stakeholders understand the implications of design decisions without placing constraints on the appearance of the control room. We encouraged the stakeholders to think outside the box for design solutions. Once the stakeholders realised that all brainstormed ideas were of value, the process was generally fun and resulted in good design suggestions.

Some of the visual ergonomics issues that we canvassed during this process included lighting that enables good visual comfort and function, the number and location of visual displays to promote good physical posture and ease of viewing, and lines of sight within the control room to facilitate good working relationships. If the very early schematic design was subsequently adopted by the business, then other professionals (e.g. lighting designers, architects, and builders) were contracted to interpret the very early schematic design, create an architectural and lighting design to meet the brief, and then build the facility.

GOOD INTENTIONS DON'T ALWAYS GO TO PLAN

Some of the early schematic design projects with smaller size clients progressed seamlessly to a built solution. This was encouraging. Instead of retrofitting solutions to existing control rooms, I was seeing built solutions that addressed visual ergonomics problems. I was making a difference! However, good intentions don't always go to plan, especially with larger size clients. Sometimes, we, meaning the control room architect, another ergonomist, and me, found ourselves enmeshed among layers of project managers and subcontractors. All were involved in the design and build of the facility, but each had their own agenda and opinion.

In one case, the client wanted us to collaborate with the fit-out team to ensure the very early schematic design was translated into the built design. However, the contractor agreement was structured to keep our working team separate from the fit-out team. The project manager did not appreciate the importance of user engagement and the ergonomic design process and expected that the control room and visual displays would be arranged in a pre-conceived way because "that is what a control room should look like".

Unfortunately, this was different to the requirements articulated by the user groups. I found this frustrating because we had spent many hours discovering the needs of the various stakeholders who were very clear about what would and would not work for them in their work environment. In addition to that, the design envisaged by the project manager had several flaws, including workstations located in positions that did not enable good lines of sight between operators or good lines of sight to a large wall-mounted display. These design flaws would make the work of the operators more difficult and uncomfortable.

Although visual ergonomics advice was contained within a report underpinning the design rationale, the advice was not adopted by the lighting designer or the built-design architect. The latter was the person who created an architectural design based on our schematic design and oversaw the construction of the control room. There was also poor communication with the built-design architect, and no communication with the lighting designer during the design and build process.

Subsequently, the visual ergonomics elements incorporated into the built product were contrary to visual ergonomics best practice. For example, we specified lighting that would not be a direct glare source for operators when they viewed the large wall-mounted display or communicated with other colleagues in the room. Also, we specified lighting that would not cause specular, mirror-like reflections on displays, obscuring critical information that operators needed to see. The built product did not meet these requirements. The ceiling luminaires were a glare source when operators looked up from their console and were visible as reflections on computer displays, making the displays unreadable. The end users were dissatisfied with the general lighting in the control room, and they switched it off to manage these visual ergonomics problems. This was such a disappointing outcome, especially when considering these problems were identified before the facility was built.

WHEN PLANS GO WELL, IT'S MAGIC!

On a more optimistic note, I have observed that success is more likely when there is good collaboration among the various design entities. In another project with similar objectives, the control room architect, another ergonomist, and I had a good working relationship with the built-design architect. The latter attended the design workshops and contributed to the discussions when required. His role in the project included contracting and liaising with a lighting designer.

I provided visual ergonomics advice as part of a larger report that described the rationale for the early schematic design. The report was read by the built-design architect who translated our recommendations to an architectural design. During the build stage, the built-design architect oversaw the work of the lighting designer. Although I didn't have the opportunity to work closely with the lighting designer, the built-design architect facilitated a discussion between me and the lighting designer about lighting products that would meet the design brief. The challenge with this project was a relatively low ceiling height in the control room that made it tricky to illuminate without causing reflected glare on the desktop or wall-mounted displays. Fortunately, the lighting designer was able to source appropriate products, and the visual ergonomics elements and lighting in the built design were reported comfortable by the end users.

SUCCESS VERSUS FAILURE

COMMUNICATION

In the first case, I described my attempts to influence the visual design of computer interfaces. I was coached on how to best communicate with the plant engineers by an on-site ergonomist who worked closely with the engineers. She understood which strategies were best to capture the attention of the engineers and explained how I should present information to capture the engineer's attention. She taught me the language of the engineers.

In the second case, I described my fortune in being able to realise my goal to eliminate visual ergonomics problems in the physical design of control rooms. I think this was successful because I was engaged by a control room architect who was familiar with the design and build process and could navigate the contractual process and speak the built-design language with the various stakeholders.

To date, I have developed skills which allow me to predict potential visual ergonomics problems by looking at two-dimensional architectural drawings and visualising what the design will look like in a three-dimensional space. However, I have not mastered the language to converse with builders to prevent visual ergonomics problems in a built product. Working with the control room architect gave me the ability to have my knowledge translated into built-design language.

I also think that there is an element of respect and trust when colleagues from the same profession converse with one another. This can help lubricate communication channels. In the second case, the control room architect and the built-design architect had worked with each other on a previous project and enjoyed shared memories of events that had occurred within their profession. This smoothed the way for a good working relationship. It would have taken a lot more time, extending beyond the project timeline, for me or the other ergonomist to build this type of rapport.

Respect and trust of fellow professionals might also explain the difficulties I described in the first case about display interfaces. I do not have a computer science or software development background, meaning I do not have the ability to fluently converse in "computer-speak". I accept this limitation. However, the engineers employed by my clients did have computer science backgrounds and were fluent in computer-speak. Still, they could not gain traction in conversations with the software developers. Admittedly, there might have been genuine technical reasons for the software developers not being able to modify the software base products. Nonetheless, should the engineers requesting the changes have also been software developers, I wonder whether the conversations and implementation of software changes would have been more successful.

EGOS

Of course, when we work in multidisciplinary teams, egos are likely to rear their head. This is even more likely when the discussions include challenging another professional about their design ideas. I think that "ego" was a strong factor in the case of the visual design of display interfaces. When challenging the engineers to modify the interface design, the software developers argued, "No-one else has complained. Why are you telling us that we are wrong?" It was almost certainly a factor in the second case when developing an early schematic design for the large-size company wanting to build a control room. There were stakeholders keen to make their stamp on the design, even if it was contrary to the wishes of the operators who were ultimately going to occupy the space.

Possibly, in working on the design of control rooms, one would think that I had developed close working relationships with lighting designers. Unfortunately, this was not the case. Although we asked our control room clients to invite the project lighting designer to ergonomics workshops with the stakeholders, very few lighting designers took up the opportunity. This did not always have an adverse effect. Some lighting designers provided elegant lighting solutions that met the user requirements and worked well in the control room. On the other hand, we also witnessed some interesting interpretations of the lighting design requirements, usually where the designer was intent on expressing their artistic flair (e.g. pendant luminaires that obscured the operator's view of wall-mounted displays, and linear extrusion ceiling luminaires that caused specular reflections on wall-mounted displays). In these cases, the design solution was detrimental to a visually comfortable control room.

At a professional level, the lighting industry recognises that multidisciplinary collaboration can be an integral component of a built design. The American-based Illuminating Engineering Society has hosted professional development forums to discuss collaborative opportunities for lighting designers. They recently invited me to write a blog for their website outlining my control room work and the advantages of multidisciplinary collaboration (Long, 2020). However, I have not experienced a good collaborative relationship with a lighting designer on a control room project, except for the instance described in the second case. In that project, the built-design architect was the mediator in a conversation between me and the lighting designer.

I was puzzled. How could I work with lighting designers? What is the barrier to a more fruitful engagement? To help me understand this, I chatted with a lighting designer recommended to me by a colleague. He candidly informed me that he would prefer to gather his information for the design brief in the way that he was accustomed. This preference would not change even if he was provided with a design brief already prepared by me and my colleagues based on the results of the ergonomics workshops with stakeholders. When I asked, "Should I do something different which would promote collaboration with lighting designers?", he disclosed that he would find it intimidating being in a room with people like me who had multiple academic degrees. I suppose that he was trying to tell me in a polite way that he thought I was an academic boffin that did not know what I was doing. As such, he would prefer to do the work himself. I know this is only one designer's opinion, and that I might gain other insights if I canvassed the opinions of other lighting designers. Nevertheless, it gave me a pause. Ergonomists frequently have undergraduate and postgraduate qualifications, and some have doctorate degrees. Is our high level of education a potential barrier to multidisciplinary collaboration? Or do we need to find more accessible language and approaches to gain the trust of our design colleagues?

CONCLUSION

I have been on an interesting quest during the past 30 years. My inability to solve the vision problems of my patients in the optometry consultation room led me to a career in ergonomics where I have endeavoured to make a difference in people's lives by improving the visual design of display interfaces and improving the physical design of control rooms.

My belief that good design will ameliorate many of the visual symptoms that people experience at work has been vindicated. I have been fortunate to contribute evidence-based visual ergonomics advice in control room projects. This has helped to create successful designs where the end users were satisfied and visually comfortable in their work environment. I wish I was more successful in influencing display interface design. Still, I quietly hope that my recommendations contained in ergonomics reports will be adopted in future iterations of software.

If I were giving advice to my younger self, then I would emphasise the importance of good communication with colleagues and stakeholders. If you cannot speak the professional language of the stakeholders, then find someone who can teach you the language or someone who can translate what you want to say into words that the stakeholders can understand. In my work with computer interfaces, I didn't master computer-speak. I believe that this was a barrier for enabling improvements to interface design. However, in my work with control rooms, I was taught engineering-speak by colleagues. I worked with an architect who could translate my visual ergonomics advice into words and images that other professionals working in the built-design environment could understand.

My second piece of advice to my younger self would be to reassure her that differences of opinion are a normal part of professional life. What matters is how you react and try to manage the differences. Unfortunately, some professional relationships are binary, win or lose. It effectively comes down to the question, "Did the good or bad design win?" However, it does not have to be that way. In our ergonomics workshops with stakeholders, we brainstormed ideas, and all ideas were respected, noted, and discussed. Once participants realised this, the process became fun and productive. Together, we generated design ideas that went beyond what we envisaged at the start of the workshop. Maybe, this is a lesson that we can take with us to our real-world interactions with other people: diverse ideas from different professionals may lead to conflict but, if well-managed, can result in robust practices and innovation.

REFERENCES

Burgess-Limerick, R. (2018). Participatory ergonomics: Evidence and implementation lessons. *Applied Ergonomics, 68*, 289–293.

Burgess-Limerick, R., Mon-Williams, M., & Coppard, V. (2000). Visual display height. *Human Factors, 42*(1), 140–150.

Hollifield, B., Oliver, D., Nimmo, I., & Habibi, E. (2008). *The High Performance HMI Handbook* (1st ed.). PAS.

ISO. (1999). *ISO 9355-2: Ergonomic requirements for the design of displays and control actuators - Part 2: Displays*.

Long, J. (2019). Prescribing for a computer user. In M. Rosenfield, E. Lee, & D. Goodwin (Eds.), *Clinical Cases in Eye Care* (pp. 40–43). Wolters Kluwer.

Long, J. (2020). Unifying the art and function of light in the built environment – the collaborative roles of visual ergonomists and lighting designers. *Illuminating Engineering Society Forum for Illumination Research, Engineering and Science (FIRES)*. https://www.ies.org/fires/unifying-the-art-and-function-of-light-in-the-built-environment-the-collaborative-roles-of-visual-ergonomists-and-lighting-designers/

Long, J., Ockendon, R., & McDonald, F. (2019). *Visual Ergonomics in Control Rooms - An example of creativity in practice* IEA2018, AISC 827, Florence, Italy.

Martin, D., & Dain, S. (1988). Postural modifications of VDU operators wearing bifocal spectacles. *Applied Ergonomics, 19*(4), 293–300.

Rempel, D., Willms, K., Anshel, J., Jaschinski, W., & Sheedy, J. (2007). The effects of visual display distance on eye accommodation, head posture, and vision and neck symptoms. *Human Factors, 49,* 830–838.

Travis, D. (1991). *Effective Color Displays: Theory and Practice.* Academic Press.

Villanueva, M., Sotoyama, M., Jonai, H., Takeuchi, Y., & Saito, S. (1996). Adjustments of posture and viewing parameters of the eye to changes in the screen height of the visual display terminal. *Ergonomics, 39*(7), 933–945.

5 Opportunities and Challenges for Designing Quality Work in Residential Aged Care

Valerie O'Keeffe
Flinders University

CONTENTS

Residential aged care services face unrelenting pressures from a growing ageing population, workforce shortages, and rising costs of care, impacting care workers' health and safety and capacity for quality care (Hodgkin et al., 2017). Incidents related to worker and client safety and service quality (i.e. errors and missed care) arise from the design of work and can be minimised through systematic work analysis and redesign (Carayon et al., 2006). Interventions are most effective when they address safety and service quality problems at the source, such as high workload, staff shortages, and inadequate resources (Oakman et al., 2019). The intervention case titled *Designed with Care* (DWC) arose from concerns from the work health and safety (WHS) regulator about high rates of work-related musculoskeletal disorders (WMSDs) and psychological injuries affecting care workers across the aged care industry. The goal was to reduce physical and psychological injury by trialling a work design intervention in an organisation to tackle upstream risk factors.

SETTING THE SCENE

GOOD WORK DESIGN: MATCHING DEMANDS WITH RESOURCES

Good work design (GWD) recognises that interventions are most successful when they are organisationally directed and systems-focused, considering dependencies between people, tools, and their physical and social environments (Karanikas et al., 2021). Effective work design matches work demands (i.e. features of work) with resources (i.e. aspects supporting the achievement of work goals) to meet those demands. Job resources can make it possible to deal with job demands and/or promote personal growth and development (Demerouti et al., 2001). Mismatched demands and resources lead to energy depletion, a situation that erodes health and produces poorer quality work outcomes (Demerouti et al., 2001).

Demands include the physical and mental content of work (i.e. its variety and complexity), workload (i.e. the effort required), exposure to the physical environment (e.g. temperature, noise, air quality), and work organisation (e.g. time pressure and hours/timing of work). Organisational resources include job control, skill development and utilisation, participative decision-making, and task variety, while social resources include support from supervisors, colleagues, friends, and family. Where job demands are high, the impact of resources becomes more critical to sustaining performance (Tims et al., 2013). Features of good work include having task variety, manageable workloads, feedback, participative decision-making, skill use and autonomy, supervisor and co-worker support, and fair remuneration (Clarke, 2015).

GWD also motivates individuals to improve their own jobs by crafting solutions to match challenging demands and available resources. It begins with a problem and/or an opportunity to improve work. In the DWC case, WMSDs and psychological injuries were the problems to solve, and better client care and worker job satisfaction were the opportunities to realise. GWD is an iterative and participative three-stage process of *Discovery, Design,* and *Realisation.* Discovery involves understanding problems and/or opportunities through engaging people affected by the work and examining the work itself. Design involves problem-solving, identifying solutions and strategies to model or trial options, and preparing for change. Realisation involves implementing solutions, innovating, and improving performance (Karanikas et al., 2021).

JOB CRAFTING

The DWC intervention aimed to optimise the balance between work demands and resources through changing work structures at the system level. Job crafting represents the proactive changes workers make to design their work to initiate positive outcomes, and it promotes engagement, job satisfaction, and resilience. Job crafting focuses on workers' opportunities to customise and actively modify their tasks and interactions with others. Tasks can be expanded or reduced in scope, the nature of relationships with others can change, or perceptions of work can be reappraised to view tasks collectively as having a broad purpose or goal (Berg et al., 2008).

Job crafting requires high levels of work engagement, evident as an active, positive work-related state of vigour, dedication, and absorption (Bakker, 2011). Co-worker

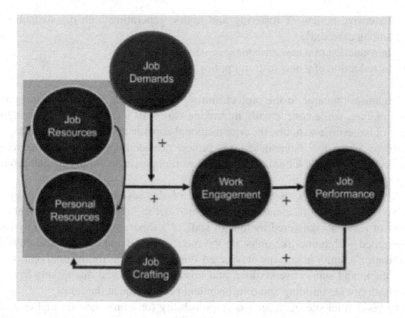

FIGURE 5.1   Model of work engagement incorporating job crafting (Based on Bakker & Demerouti, 2008.)

support, performance feedback, skill variety, autonomy, and learning opportunities promote work engagement, where workers harness motivation to adjust demands, resources, or both. Job crafting becomes even more significant when demands are inherent in the job itself, as in aged care, and not easily modified (Cooley & Yovanoff, 1996). Given the significant recent restructuring in the host aged care organisation of this case, the DWC intervention focused on increasing care staff resources through job crafting, as depicted in Figure 5.1. An intervention bolstering personal and job resources was expected to increase work engagement and improve job performance by increasing capacity to respond more effectively to high job demands.

THE CONTEXT AND ITS ACTORS

The mission of the aged care provider involved in this project is to be a leader in providing quality residential care for aging clients based on Christian principles that maintain the dignity, rights, and values of the individual. Organisational values also reflect respect for people and their right to live in a home-like environment. The organisation aims to provide best-practice person-centred care delivered by highly trained staff with adequate resources and time to fulfil their roles effectively. Following a period of organisational change and restructure, the organisation experienced the following:

- Restructuring of roles and responsibilities, with significant changes in key personnel in the Head Office and onsite;

- Extensive review of rostering and hours, generating high dissatisfaction among care staff;
- Introduction of a new case management model; and
- Introduction of a new restraint policy.

The organisation came to the project aiming to embed the new case management model of care, while concurrently improving care staff's job quality. Following the review of rostering and hours, the organisational appetite to change structural aspects of care work was low. Nonetheless, the project provided opportunities to adopt the new model as a vehicle for enhancing direct worker participation in care and promoting job satisfaction.

Wisteria Gardens (a pseudonym) is one of the several sites operated by the involved not-for-profit residential aged care provider in Australia. It provides services to tens of residents delivered by tens of staff in a modern, dementia-friendly facility designed with home-like units, on the metropolitan fringe of a capital city. The organisation's senior management selected Wisteria Gardens for the intervention due to its stable and supportive staffing, and minimal anticipated disruptions from the organisation-wide building and refurbishment programme at the time.

The residential site manager has responsibility for administration and customer services, WHS, hotel and property services, clinical care, continuous improvement, and lifestyle and pastoral care. The clinical nurse manager supervises four clinical leaders who ensure high standards of clinical care through their supervisory role of direct care workers, the latter including enrolled nurses and personal care attendants. The new clinical leader role was the centrepiece of the intervention, seeking to expand their supervisory skills and strengthen care worker contributions to improving care.

My role as an ergonomist was to be a knowledgeable facilitator. Trained as an allied health professional, I brought familiarity with the health and well-being issues of eldercare, allowing shared language, and understanding of current work practices from care staff perspectives. I facilitated a co-design process where problems were structured and solutions were identified collaboratively with care staff and management. Ideally, a co-design process would include consumers of care in framing problems and opportunities. However, management did not include consumer perspectives as they considered the intervention primarily a WHS initiative.

Indeed, the project premised that risks are inherent in the work itself and improving work design would be beneficial for reducing worker injuries and enhancing client care outcomes. However, in the six months' time between applying for funding and commencing the project, the provider experienced extensive change, including the loss of the project sponsor. The organisation was change-weary, leading to difficulty in gaining new management's trust and commitment. They were reluctant to embark on further change, concerned that issues arising from the previous change would be reignited.

Challenges also arose in gaining care workers' trust and confidence in the intervention. Recent reviews of hours and rosters left care staff cynical that positive change could be realised, and they were feeling little enthusiasm for participating in a project-steering group since they were time-poor and fatigued. Strategies to build commitment and participation included identifying workers of influence, providing them with information to promote the project, and fostering grassroots optimism that small-scale changes could be achieved.

THE DISCOVERY JOURNEY

Since the problem was initially related to injury occurrences, WHS data could inform the extent and nature of injuries. The WHS regulator provided data describing occurrences of WMSD and psychological injuries across the state's aged care industry. We compared these with data from all worksites of the aged care provider, including Wisteria Gardens. Data were collected for the four years before commencing the project and were evaluated as percentages of total claims. Over that period, the organisation experienced 310 accepted workers' compensation claims, with 23 (7.4%) occurring at Wisteria Gardens, showing the site performed slightly better than most other sites with about 9% of the total claims each.

The organisation experienced a higher proportion of claims for WMSD (49.7%) than the industry overall (44.5%), but performed better with lower psychological claims than the industry (2.6% vs 4.4%). However, this could suggest the possibility of reluctance to report. I noted that at Wisteria Gardens, they collected injury and incident data for all classifications of potential injury events (i.e. report only, first aid only, and medical treatment claims) and included injuries directly resulting from client behaviours. Collectively, these data provided a baseline for designing and evaluating solutions.

Injury data draw a picture of *what* injuries are occurring, *when, where,* and to *whom,* but are less informative about *how* injuries arise during work. Interviews and surveys support a more holistic assessment of the interactive nature of risks, accounting for the job in its whole social and organisational context. These methods identify upstream factors less evident in task-based analyses of technical risk factors. Surveys and interviews also enable participation and expressions of diverse knowledge, bringing rich insights from those with first-hand experience. My clinical experience had armed me with intimate knowledge of how care work is constructed, and that it depends on effective teamwork between colleagues and clients. Interviews are critical in understanding and designing good work because they tap that tacit knowledge and dynamic interplay between people and tasks and reveal how staff flexibly respond to needs, wants, and challenges.

Fourteen care staff from all units participated in the discovery phase interviews. Interviewing care staff during work time is challenging as staff are time-pressured, and it is difficult to relieve them for around 30 minutes without interruption. Also, influenced by perceived cost pressures, releasing care workers to take part in interviews required ongoing negotiation with management, despite the funding proposal budgeting for reimbursing their time. Strategies to counter these challenges included engaging management regularly in developments, emphasising and paying close attention to enacting the co-design process and principles, and responding promptly to questions and concerns using the best evidence. As an external facilitator, I regularly provided examples from other workplaces in aged care and broader industry to explain the rationale for how improvements in work design would benefit care.

Having had experience in aged care and broader industry helped to build credibility with the Wisteria Gardens team, and confidence I would bring new perspectives and question the status quo.

During the interviews, I invited participants to describe their work, what they liked most and least, the greatest demands and priorities, risks to their health and safety, their involvement in decision-making, and the characteristics of workers most valued in the facility. I also provided participants with a summary of the injury statistics analysis and evidence-based information about the shared causative mechanisms for WMSD and psychological injuries. I prompted them to reflect on this information and identify simple changes that they believed could improve their work while maintaining service delivery.

The interviews revealed workers did not feel adequately skilled or supported when dealing with the challenging behaviours of some residents, and the impact of workload made it difficult for them to be 'present' and provide individualised care. Staff also felt that the residents' right to a home-like environment was prioritised above the workers' right to a safe workplace. The carers also highlighted concerns about manual handling when responding to residents with challenging behaviours, given the recent changes in the restraint policy. Consistent with best practices in aged care policy, the review of the restraint policy led to bed rails being eliminated, and a gradual introduction of new beds and associated manual handling equipment, including hoists, bed poles, and sensor mats. Overall, the interview findings emphasised the busyness of the work environment, the challenge of communicating and coordinating work, and the desire of care staff for greater support.

Additionally, I administered anonymous surveys seeking care workers' perceptions of work design factors to assess the risks and opportunities for improving care work. Surveys examined four constructs involving qualitative and quantitative variables: body part discomfort (Dawson et al., 2009), job demands and resources (Karasek et al., 1998), work engagement (Seppälä et al., 2009), and psychosocial safety climate (PSC) (Hall et al., 2010). I distributed the surveys in online and hard copy formats through staff emails and mailboxes, in the staff room, and through direct personal invitations. The responses to the survey highlighted that carers experienced considerable levels of mental and physical fatigue and localised body part discomfort, plus high physical and mental work demands.

Furthermore, to gain a nuanced understanding of the organisational culture and practices, I attended the site one day each week during the nine months of the project. Work occurs within a context that reflects organisational values, and successful interventions must consider that context (Zadow et al., 2017). My purpose was to shadow staff during daily activities across morning and afternoon shifts. This built relationships and trust, enabled observations of work practices and interactions between staff and with residents, and offered an opportunity to review organisational documentation (e.g. enterprise agreements, job descriptions, annual reports, policies and procedures, and training material).

The organisational scan revealed a comprehensive manual handling programme in place, including regular reviews of controls, integration with care planning, maintenance of equipment, and targeted training. The organisation had also implemented an industry award-winning dementia care programme built on honouring residents' dignity, choice, and individuality. Following the recent direct care staff rostering and hours review, direct contact for care at Wisteria Gardens was on average 3.3 hours/day and per resident. The industry standard is 2.84 hours/day. Comparable to Australia, Canadian standards require 3.0 hours per resident per day of direct care (Eagar et al., 2019), while research on missed care suggests an average of 4.3 hours/day is necessary to provide quality care (Willis et al., 2016). Hence, Wisteria Gardens was sitting somewhere between the industry average and evidenced standard of quality care provision.

Combined, the findings from the interviews, surveys, and the organisational scan highlighted the potential value of increasing structured communication and information-sharing to facilitate planning and problem-solving in a time-pressured environment. Providing quality aged care services is challenging, given the expectation to respond to increasing service demands, and the growing complexity and individualised care needs of residents. These factors intensify the physical and psychological demands placed on care staff. Considering management reluctance to revisit recent modifications to rostering and staffing, and the systemic sources of these risks, large-scale work design changes were deemed unfeasible. Yet, this project provided an opportunity to highlight the value of small-scale localised approaches to improving work and its outcomes.

THE CO-DESIGN JOURNEY

Designing solutions is a recursive process requiring imagination, creativity, and innovation stimulated through active participation by those affected by the work. Review and integration of evidence gathered in the discovery process must also inform design. It is worth remembering that solutions often need to meet various criteria beyond addressing the central problem, acknowledging that some demands are inherent in the work and not easily altered without large-scale transformation. The DWC intervention had to be small-scale, benefit care, not incur organisational costs, and improve job resources to reduce risks of WMSDs and psychological injuries.

Indeed, the process of developing solutions to an implementable state often identifies gaps in knowledge and necessitates further discovery to collect fresh evidence and inform how to embody and apply a solution. Some ideas may innovate and advance another idea, so all ideas are potentially rich sources of insight and should not be hastily discarded. My experiences working in engineering sciences to implement technology in production processes have shown me the value of studying how other industries have solved similar problems and testing and trialling tools and techniques at a micro-scale as proof of concept. Small wins garner support and motivation to try further improvement ideas, building momentum for GWD.

As workplace participation is essential for designing safe work (Pehkhonen et al., 2009), we adopted a co-design approach. To be sustainable, job design improvements should concentrate on psychological risks and WMSD prevention and improve standards of care and overall business performance. Cultivating improved communication among care staff as a job resource was chosen as the focus of the intervention. Improved communication is effective in modifying risks for WMSDs and psychological harm arising from the high physical and psychological demands (Wagner et al., 2015).

Traditional documented communications often fail. They can be time-consuming and cannot readily capture the dynamism of aged care work because employees are less inclined to document subtle changes. Formal reporting processes often miss much of the fluid nature of real-time interpersonal exchanges that support spontaneous problem identification, generate solutions, and share ideas. The reality of care work involves tensions between time pressures and responses to individual client needs. Dynamic approaches are timely, are responsive, and support 'making sense' of clients' cues, organisational goals, and staff safety and well-being (O'Keeffe et al., 2015). Thus, care staff could benefit from interactions through sharing experiences and collectively 'make sense' of these in the context of their work.

As focus groups are valuable mechanisms for generating ideas and selecting solutions, I facilitated separate focus groups involving management and a representative sample of eight care staff to reflect on the discovery phase findings and feasible solutions. Both groups agreed to trial an intervention introducing structured communication tools to underpin more effective and responsive problem-solving and information-sharing during delivery of care.

The co-designed intervention was a structured problem-solving approach and we named SHARE4CARE that could be used in various settings in the aged care work environment. It was inspired by the tool STOP and WATCH (Ouslander et al., 2014) and the Team Strategies and Tools to Enhance Performance and Patient Safety (TeamSTEPPS) programme (Guimond, 2009). The SHARE4CARE acronym summarises the communication process and skills required for ad hoc 'on-the-go' problem-solving:

S = Stop, speak
H = Hear (listen)
A = Acknowledge or ask more questions
R = Reflect
E = Engage together
4 = 4 minutes – the time that the process should take
C = Coordinate
A = Act
R = Review effectiveness
E = Embed improvements into daily practice

SHARE4CARE provided a framework for spontaneous problem-solving and communication between two or more carers. The tool encouraged staff to come together around a specific issue to be resolved, resulting in a shared plan of action and review. SHARE4CARE also enabled staff to express observations and experiences as inherent in giving care, rather than a collection of technical hazards stripped of their context. This way, we tackled WMSDs and psychosocial risk factors upstream through work design and the social context where hazards were experienced, rather than constraining our views on downstream factors like biomechanical risks.

Additionally, Buzz meetings were the idea of site management to increase staff participation in care planning for individual residents. These were clinical leader-led 20–30 minutes meetings in each unit that allowed care staff to come together with their clinical leader around current issues affecting care and be connected to the current 'Buzz'. Meetings were ad hoc and held across morning and afternoon shifts. The clinical nurse manager's goal was to use these sessions to tap carers' insights into residents whose care plans and funding were under review. Each meeting had a topic or resident of focus, supplemented with a general discussion on what was working well or not in the unit. The SHARE4CARE tool was also used to guide discussion during Buzz meetings. Management and care staff supported implementing the SHARE4CARE communication tool and regular ad hoc Buzz meetings. Together, these interventions addressed care staff requests for improved communication, information-sharing, and supervisory support.

THE REALISATION JOURNEY

Realisation in the work design process refers to the translation of design outcomes to deliver tangible benefits for everyone affected by the work, which in this case included management, care staff, and clients. It involves working closely with workplace partners to understand the organisational context and ensure implementation and uptake can be achieved through effective planning, resource allocation, and participation. Our communication intervention aimed to enhance information and social resources to support problem-solving and optimise challenge, autonomy, skill utilisation, and learning.

I introduced the SHARE4CARE tool in a half-day workshop delivered to clinical leaders and care staff. The workshop began with briefly reviewing the discovery and design stage outcomes and then presented the SHARE4CARE communication tool. The rest of the workshop provided theory and practice in using effective communication skills (e.g. professional, factual, and assertive language, 'I' statements, managing emotion, active listening, reflection, apologising, and perspective-taking). Practice involved role-playing in pairs, using vignettes typical of care work drawn from interview examples.

After the workshop, I provided all staff with a leaflet of effective communication tips and a lanyard card outlining the steps involved in the SHARE4CARE process. Clinical leaders, enrolled nurses, and personal care workers of influence were targeted to promote and coach colleagues in using the tool. Training also included information on introducing Buzz meetings, which were the responsibility of the clinical nurse manager and clinical leaders, supported by me. SHARE4CARE was implemented immediately following the workshop, but only across morning and afternoon shifts because of limited staffing.

Unfortunately, the delivery of the training workshop had been delayed, thus significantly reducing the time available for implementation and evaluation, given the strict completion timelines specified by the funding body. Contributing to the delay were also changes of management at the Wisteria Gardens site and the absence of the senior manager site liaison for six weeks, combined with two gastroenteritis outbreaks and competing priorities for care staff training time. Consequently, the intervention was implemented for only three of the planned six months.

Evaluation is a critical component of realisation to understand whether the intervention was effective and how outcomes were achieved. Each week, I conducted spontaneous 5–10 minutes interviews with care staff and clinical leaders about experiences from using the SHARE4CARE tool and participation in Buzz meetings. Evaluation metrics included numbers of Buzz meetings, uptake of the tool, and ratings of their effectiveness. After three months, interviews and surveys were repeated to evaluate change. All staff interviewed were aware of Buzz meetings, although only 40% had attended at least one because meetings had not been implemented for some afternoon shifts or were held when individual staff were not on shift.

The interviews revealed that despite staff reporting workloads having intensified in one unit, improvements were made to equipment and teamwork, resulting in reduced WMSDs risks. Staff perceived residents had become more dependent, requiring greater teamwork to provide care. Following the commencement of the new restraint policy, specific training was also provided on handling sensor mats and using new equipment. During the intervention, large areas of carpeted flooring were also removed, reducing forceful manual handling effort.

For psychological risks, staff reported greater support from supervisors and managers, being listened to and consulted, and better co-operation. Carers reported improvements in management support when responding to challenging behaviours by residents or family members. Co-operation, teamwork, and information-sharing between clinical leaders and care staff had increased since Buzz meetings begun. Participants attributed this to greater staff awareness, listening to each other, and willingness to share information. Clinical leaders also observed improvements in care delivery, planning, and documentation due to improved information-sharing.

The findings from the surveys showed modest improvements in physical and mental fatigue and body part discomfort in the upper back and shoulders. Physical work demands improved with reduced frequency of demands, though with the same effort required. Mental demands remained similar, although qualitative comments revealed less need to remember information, make a mental effort, and show emotions inconsistent with those being felt. Job resources had a modest improvement, most notably in being able to express emotions without negative consequences, choosing how to do tasks, and being able to use skills and knowledge to solve complex tasks. Engagement scores remained unchanged post-intervention, with modest improvement for inspiration.

The most significant improvement was in the climate of psychological safety, demonstrating perceived enactment of organisational policies, practices, and procedures to protect worker health and safety. Pre-intervention Wisteria Gardens PSC score indicated a high risk for psychological injury and ill-health, which decreased to moderate levels post-intervention. Staff feedback on the quality of communication during Buzz meetings also showed improvements in feeling energised, strong, enthusiastic, inspired, and keen to go to work.

DESIGNED WITH CARE: SUCCESSES AND FAILURES

Our intervention, while implemented only for a short period, showed a modest positive impact on care staff risks for WMSD and psychological injury risk, achieving its primary aim. We experienced measurable improvements in communication quality and relationship satisfaction, which translated into tangible improvements in care delivery, planning, and documentation. A significant success was the rapid improvement in PSC for Wisteria Gardens; within three months, the increase in supervisory support was instrumental in reducing the risk profile from high to moderate risk, providing impetus for continued improvement.

The key to success in our intervention was the participative co-design process, for its value in bringing participants together with the common purpose of improving the quality of work. The co-design process was a success and a failure at the same time, given that the extent of participation was limited. Nonetheless, persistence, enthusiasm, and creative approaches to engagement achieved quality participation and were the key to achieving positive workplace improvements.

Furthermore, successful outcomes of the DWC intervention were more readily achievable because we used the job crafting model to guide the design and content of the intervention and evaluation. The model was valuable in highlighting the range of measures required to demonstrate and explain the holistic change and provided a structure for describing the intervention to participants. It helped build participants' confidence that the intervention was plausible and would be worthy of investing their effort. Using mixed methods also contributed to demonstrating tangible improvements, where qualitative data supplemented quantitative ratings to illustrate specific examples of improved practice and experiences of more collaborative interactions.

On the other hand, the low engagement, commitment, and trust of senior management were significant barriers to undertaking the intervention. Being an external consultant, this manifested as limited access to the site for one day/week, and always having a Head Office liaison person on site. The result was low participation and repeated delays to implementation. In the DWC project, half the number of staff required were released to attend training, limiting exposure to, and uptake of the intervention. To counter this, I invested heavily in building trust in relationships with senior management at Head Office and Wisteria Gardens by engaging with them regularly, checking in on current happenings, offering suggestions on strategies, and providing information on successful interventions. I also maintained high visibility with care staff through informal interactions during work time and breaks.

Likewise, I identified and met informally with individuals and small groups of curious care staff, briefly explaining the intervention purpose, the SHARE4CARE tool and Buzz meetings, engaging staff, and disseminating the intervention at the grassroots level. A consequence of low participation was the small sample size for surveys, interviews, and a low uptake of the SHARE4CARE training and tool use. This meant that the results were not conclusive overall despite them being positive and promising.

THE LESSONS LEARNED

Intervention research is dynamic, and predicting the course and outcomes of change is challenging and surprising. The lessons learned from the DWC project may be instructive for other work design interventions.

First, there is never an ideal time to implement change since there will always be competing priorities. Respond to the immediate needs by starting small with one change, in one workgroup if necessary. Use these experiences to build foundations for pursuing more ambitious problems and opportunities. Build trust and embrace a co-design approach to gain commitment and learn from participants' experiences.

Second, change can be achieved quickly when starting small, but sustainable change takes sustained effort. Impart technical and social skills to others, especially leaders and workers of influence, and work at achieving integration with other organisational processes and priorities. Most importantly, be flexible and adaptive and sell the merits of the intervention and value of work design.

Third, implementing successful organisational interventions requires ongoing demonstrable commitment from senior and middle management. Their role is critical in accessing staff, providing positive messaging, prioritising intervention activities, and generating momentum. Hence, invest in achieving commitment at all levels of the organisation. Adopting an organisational focus, even when working in a localised group, maximises the potential for sustainability and success through integrating changes in the organisation's practices and culture. The facilitator must conscientiously support management to lead the change, creating a symbiotic relationship of mutual value. Sustaining outcomes requires maintenance of new behaviours after the facilitator has left, so the support of senior management and workers of influence is critical to success.

Fourth, we must build the evaluation into the intervention from the discovery stage and throughout design and realisation. Using a model or theory to guide intervention design and implementation helps explain the purpose and process to participants and promotes robust evaluation. Reactions to change can vary and may occur quickly but not necessarily be sustained. Also, be mindful that social desirability may influence reporting of change, where participants provide responses perceived to be more acceptable to the facilitator or management.

Finally, an effective facilitator must maintain persistence and optimism, transmitting it to intervention participants, catalysing enthusiasm and uptake. Effective planning is a key to guiding project success, but the facilitator must be prepared to adapt and be flexible, given busy workplaces have competing priorities. Honouring the co-design process throughout project realisation will ensure activities are relevant, expectations remain realistic, and problem-solving and review are ongoing, consistent with the contract of engagement.

REFERENCES

Bakker, A. B. (2011). An evidence-based model of work engagement. *Current Directions in Psychological Science*, *20*(4), 265–269. DOI: 10.1177/0963721411414534

Bakker, A. B. & Demerouti, E. (2008). Towards a model of work engagement. *Career Development International*, *13*(3), 209–223. DOI: 10.1108/13620430810870476

Berg, J. M., Dutton, J. E., & Wrzesniewski, A. (2008). What is job crafting and why does it matter. Retrieved from https://www.researchgate.net/publication/266094577 on 21 December 2021.

Carayon, P. A. S. H., Hundt, A. S., Karsh, B. T., Gurses, A. P., Alvarado, C. J., Smith, M., & Brennan, P. F. (2006). Work system design for patient safety: The SEIPS model. *BMJ Quality & Safety*, *15*(suppl 1), i50–i58. DOI: 10.1136/qshc.2005.015842

Clarke, M. (2015). To what extent a "bad" job? Employee perceptions of job quality in community aged care. *Employee Relations*, *37*(2), 192–208. DOI: 10.1108/ER-11-2013-0169

Cooley, E., & Yovanoff, P. (1996). Supporting professionals-at-risk: Evaluating interventions to reduce burnout and improve retention of special educators. *Exceptional Children*, *62*(4), 336–355. DOI: 10.1177/001440299606200404

Dawson, A. P., Steele, E. J., Hodges, P. W., & Stewart, S. (2009). Development and test–retest reliability of an extended version of the Nordic Musculoskeletal Questionnaire (NMQ-E): A screening instrument for musculoskeletal pain. *The Journal of Pain*, *10*(5), 517–526. DOI: 10.1016/j.jpain.2008.11.008

Demerouti, E., Bakker, A. B., Nachreiner, F., & Schaufeli, W. B. (2001). The job demands-resources model of burnout. *Journal of Applied Psychology*, *86*(3), 499. DOI: 10.1037/0021-9010.86.3.499

Eagar, K., Westera, A., Snoek, M., Kobel, C., Loggie, C., & Gordon, R. (2019). *How Australian residential aged care staffing levels compare with international and national benchmarks*. Centre for Health Service Development, Australian Health Services Research Institute, University of Wollongong.

Guimond, M. E., Sole, M. L., & Salas, E. (2009). TeamSTEPPS. *The American Journal of Nursing*, *109*(11), 66–68. DOI: 10.1097/01.NAJ.0000363359.84377.27

Hall, G. B., Dollard, M. F., & Coward, J. (2010). Psychosocial safety climate: Development of the PSC-12. *International Journal of Stress Management*, *17*(4), 353. DOI: 10.1037/a0021320

Hodgkin, S., Warburton, J., Savy, P., & Moore, M. (2017). Workforce crisis in residential aged care: Insights from rural, older workers. *Australian Journal of Public Administration*, *76*(1), 93–105. DOI: 10.1111/1467-8500.12204

Karanikas, N., Pazell, S., Wright, A., & Crawford, E. (2021). The What, Why and How of Good Work Design: The Perspective of the Human Factors and Ergonomics Society of Australia. In: Rebelo, F. (eds). *Advances in ergonomics in design. AHFE 2021. Lecture Notes in Networks and Systems*, vol 261, pp. 904–911. Springer, Cham. DOI: 10.1007/978-3-030-79760-7_108

Karasek, R., Brisson, C., Kawakami, N., Houtman, I., Bongers, P., & Amick, B. (1998). The Job Content Questionnaire (JCQ): An instrument for internationally comparative assessments of psychosocial job characteristics. *Journal of Occupational Health Psychology*, *3*(4), 322. DOI: 10.1037//1076-8998.3.4.322

O'Keeffe, V. J., Tuckey, M. R., & Naweed, A. (2015). Whose safety? Flexible risk assessment boundaries balance nurse safety with patient care. *Safety Science*, *76*, 111–120. DOI: 10.1016/j.ssci.2015.02.024

Oakman, J., Macdonald, W., & Kinsman, N. (2019). Barriers to more effective prevention of work-related musculoskeletal and mental health disorders. *Applied Ergonomics*, *75*, 184–192. DOI: 10.1016/j.apergo.2018.10.007

Ouslander, J. G., Bonner, A., Herndon, L., & Shutes, J. (2014). The Interventions to Reduce Acute Care Transfers (INTERACT) quality improvement program: An overview for medical directors and primary care clinicians in long term care. *Journal of the American Medical Directors Association*, *15*(3), 162–170. DOI: 10.1016/j.jamda.2013.12.005

Pehkonen, I., Takala, E.P., Ketola, R., Viikarai-Juntura, Leino-Arjas, P., Hopsu, L., & Riihimaki, H. (2009). Evaluation of a participatory ergonomics intervention process in kitchen work. *Applied Ergonomics*, *40*(1), 115–123. DOI: 10.1016/j.apergo.2008.01.006

Seppälä, P., Mauno, S., Feldt, T., Hakanen, J., Kinnunen, U., Tolvanen, A., & Schaufeli, W. (2009). The construct validity of the Utrecht Work Engagement Scale: Multisample and longitudinal evidence. *Journal of Happiness Studies*, *10*(4), 459–481. DOI: 10.1007/s10902-008-9100-y

Tims, M., Bakker, A. B., & Derks, D. (2013). The impact of job crafting on job demands, job resources, and well-being. *Journal of Occupational Health Psychology*, *18*(2), 230. DOI: 10.1037/a0032141

Wagner, S. L., White, M. I., Schultz, I. Z., Williams-Whitt, K., Koehn, C., Dionne, C. E., ... & Wright, M. D. (2015). Social support and supervisory quality interventions in the workplace: A stakeholder-centred best-evidence synthesis of systematic reviews on work outcomes. *The International Journal of Occupational and Environmental Medicine, 6*(4), 189. DOI: 10.15171/ijoem.2015.608

Willis, E., Price, K., Bonner, R., Henderson, J., Gibson, T., Hurley, J., Blackman, I., Toffoli, L., & Currie, T. (2016). *Meeting residents' care needs: A study of the requirement for nursing and personal care staff.* Australian Nursing and Midwifery Federation. Retrieved from https://agedcare.royalcommission.gov.au/system/files/2020-08/ANM.0001.0001.3151.pdf on 4 September 2022.

Zadow, A. J., Dollard, M. F., McLinton, S. S., Lawrence, P., & Tuckey, M. R. (2017). Psychosocial safety climate, emotional exhaustion, and work injuries in healthcare workplaces. *Stress and Health, 33*(5), 558–569. DOI: 10.1002/smi.2740

6 When Success Is Not Success, We Strive to Do Better

Sara Pazell
ViVA health at work

CONTENTS

When someone asks me about what I do for a living, I hesitate. How can you quickly, in a sound byte or a single phrase, translate the complexity of human factors and ergonomics that draws upon environmental, engineering, psychology, social, organisational, kinesiology, and exercise sciences? The design practices have health, well-being, and sustainability aspirations also. My answer may be simply,

Sara: *"I am a work design strategist. I help design work for health and productivity"*. This is usually met with a quizzical response and a remark like,

Curious others: "Oooohhhh. Uh-huh". If the person dares to venture further, I might be asked, "So, do you work in human resources?"

Sara: *"Well, we can leverage the workforce strategy to inform design. We design for diversity to help enact inclusivity policies, so the human resources business units could have some of their objectives met by what we facilitate in work design"*.

Curious others: "Um, okay, so do you work in health and safety?"

Sara: *"Some ergonomists align their work in the domain of health and safety. I prefer to anchor my work to design that can address health and safety concerns with productivity, workforce strategy, well-being, employee engagement, procurement, operations, engineering, sustainability, continuous improvement, facilities management, technology adoption, and similar. In other words, work design strategy extends and unites all business units, if permitted, with an overarching design philosophy about human performance"*.

DOI: 10.1201/9781003349976-6

Curious others: "Okay, so you mentioned work as an ergonomist, like agriculture?"

Sara: *"Ah, almost. That is an agronomist, and sometimes it feels like we push stuff uphill; this is true, but ergonomics is a unique discipline".*

Curious others: "Oh! Ergonomics. So, you help people with their office workstation desks, chairs, and computers, right?"

Sara: *"Well, workstation ergonomics is one area of our work. Ergonomics and human factors address how people think, socialise, move, act, interface with equipment, and work productively and sustainably. We help design tools, equipment, environments, and work systems across all industries, like mining, transportation, finance, education, health, manufacturing, retail, government services, and, yes, agriculture. Ergonomics is not a product, like an "ergonomic mouse", it is a design practice".*

Curious others: "Wow. Errr... how confusing".

And so, it goes... When a discipline of work or a profession is poorly recognised in industry, barriers can exist. A lack of familiarity about what someone does or how they should think can lead to biases, misinterpreted expectations or assumptions, and the imposition of and constraints about the agency and scope of work. If people do not know what you do, how can they refer to your services or understand how to let you advise per your knowledge base, versus limit you to a question that needs answering per their world view?

MANUAL TASK RISK MANAGEMENT

DISCOVERY PHASE

I worked as an occupational health and ergonomics adviser while on a casual contract with a large construction organisation. This case was housed in their road construction and maintenance division. Ergonomics was new to the organisation, and the teachings were in their infancy: Their programs were immature. I spent my days observing operations and speaking with the staff to get to know them and better understand their work. I ran workshops about manual task risk management, providing interactive education about work tolerances, task design, and risk factors of musculoskeletal disorders. I also taught about general health systems and biopsychosocial models of health, including what we can do to inspire health through good nutrition, hydration, rest, social relations, and activity.

The workers in the field, who were, by and large, a male population, began to acknowledge my keen interest in their work, not as an auditor, but as someone vested in the design of work for their health and productivity. As such, they began to trust me. They became accustomed to my presence and my questions and, I believe, looked forward to some of our conversations. I was enthusiastic about my work. This was a new world for me. One of my favourite aspects of ergonomics advising was learning about how the world ticked, the mechanisation of industrialised societies, a field trip with every outing to understand "a day in the life of" a new neighbour. This intrigued me and contributed to my youthful vigour in my profession, no matter my seasoning, chronological age, or experience.

DESIGN PHASE

After one such training delivery of manual task risk management for field and plant mechanics, I encouraged participants to nominate tasks that could be more efficient. I invited them to imagine their work and picture opportunities based on tasks that were meaningful and important to them and the business. I asked them to envision what "new" might look and would feel like could the task be re-designed and unfettered by process or budget. We spoke about what they dreamed could happen, how it might be designed, and how that destiny could be shaped for ongoing resilient performance. This was an appreciative approach to work design (Bushe & Kassam, 2005).

As fantastic as this sounds, believing that change is possible in what usually might be a menial task enlivens work. A manual task may seem routine because of familiarity, which can breed complacency, but once workers are given agency to become partners in designs (Burgess-Limerick, 2018), these daily tasks are made more meaningful; they signify the importance of their work and how their imagination can be applied. Being given a voice means that you and your ideas matter, and the work design can better reflect the diverse needs of those who do those daily tasks (Pazell, 2021). One of the tasks that they nominated for re-design was removing and replacing paver wheels in preparation for tyre changes (Pazell, 2018).

An asphalt-paving machine lays, forms, and partially compacts a layer of asphalt on roadway surfaces. As with most earth-moving construction equipment, it is a large and heavy piece of machinery. In this case, the tyres used by the machinery measured 1,470 mm diameter and weighed 400 kg when ballasted and filled with water, a common practice to stabilise heavy machinery tyres. Removing the wheels to prepare for the tyre change was a three-person job. The tasks required equipment isolation; chocking the wheels; removing bolts, the side arms, and adjusting hydraulics; releasing the tyre and assembly to manoeuvre on forklift tynes; and forklift operation. The tyre was moved from the workshop to an outside storage area, and water was released from the tyre. An outside contractor changed the tyre once the wheel was set aside and emptied.

Once the wheels were chocked by blocks, one mechanic operated a forklift. There were two other mechanics involved in the tyre-changing task, and they used hand tools such as rattle guns, spanners, sockets, breaker bars, crowbars, chains, and a jack stand to remove the wheel (Figure 6.1). There were hazards in this work, and the mechanics helped identify these: working near the mobile plant, pinch points, and manual tasks with heavy aspects requiring extreme exertion and awkward body postures. There were hand tools with vibration and noise exposures too. Hence, there were risks of collision, compression injury, musculoskeletal disorder, and slips/trips/falls. Some of the explosive risks were mitigated for these workers because of the contracted work for removing and replacing the tyre. Cognitive distractions were also common in the workshop area, arising from emergent service requests, especially related to breakdowns and other urgent work, or visitors from central offices, suppliers, etc. Clutter in the workplace was not uncommon, such as scattered tools, machinery, or equipment, especially when in proximity to colleagues working closely on neighbouring tasks within the covered bay.

FIGURE 6.1   Paver wheel removal inside the forklift tynes.

The seven mechanics, including their team leader, were encouraged to participate in developing design concepts. I asked them to consider what they wanted to achieve if the task could be done differently. They asked for a more efficient task; fewer workers required to do the job; reduced risk for collision with the forklift, hand injuries, and back discomfort; and general creation of ease so that this was not a bemoaned task assignment.

Ironically, this was not the first time that this crew attempted to address the risks of this task. They had previously identified the paver wheel-changing task as a hazard on their reporting logs, but they felt that it had gotten nowhere and did not receive support or feedback. The hazard log had not escalated into a risk assessment or any further action, so they left it be, as work that simply needed doing. This, to me, indicated a measure of trust and hope that, when they raised the issue through our training forums, they believed that change might be possible.

I felt that this was achievable, mainly because of the workers' commitment. It was a well-defined task with repeatable steps and firm boundaries that might be reconstructed with the proper assistive devices. I raised the issue by speaking with a regional contracting and line manager. I noted it on my monthly management report also. I asked to escalate this for manual task risk assessment, which I was happy to conduct and oversee. A meeting was held with the regional contract manager, the capital equipment manager, and me. The capital equipment manager wanted to know why the issue was being addressed and whether the task had been reported through standard hazard reports previously. He was taken aback when it was explained that the crew members' requests had gone unnoticed for more than a year. The task and related issues had been reported in hazard logs. It was flagged for quality improvement on a register more than a year prior, but the requests had gone unnoticed.

Despite these conversations and reports, seven months transpired. The workers remained inspired mostly because I checked on them regularly and continued to highlight the outstanding issue on monthly reports received by state and regional management and central safety and operational support teams. The workers continued to examine solutions, often on their own time, including attending a supplier warehouse and other road-construction mechanical workshops. I partnered with them and sent suggestions of wheel dollies from online research material and investigations of those used in the mining industry. However, they explained that most were ill-suited because they could not fit within the narrow space of the wheel arch. This is important: as co-designers, the workers are subject matter experts and, when involved, can help determine task solutions that are likely to work (Burgess-Limerick, 2018).

Eventually, the workers found a potential solution, one used by the paver manufacturer, a unique wheel dolly that fit the wheel arch of the pavers in use. However, line management denied their request for a product trial. This changed when the state-wide general manager directed the trial. This manager visited the depot and heard directly from the maintenance crew. He learned about the task and listened to their concerns. He confronted the line and capital equipment managers. The next day, the workers advised me that they could be involved in the risk assessment with me and, if warranted after the review of my report, trial new equipment and construct meaningful change in the design of the task. However, that was coupled with a dressing down. They received a severe warning from their line management that they were not to speak about their management channels or go around their line of communication again. Apparently, this was prevalent in the culture in this work area, hence their reluctance to advance the issue beyond what had been tried through reporting registers. Had they actioned otherwise, there would be a risk of retribution, such as unfavourable work assignments, constraints on promotions, or another dressing down and ongoing conflict. However, the consequences were largely left unspoken. The imagined penalties may have been worse than reality, but the threat hung in the air, nonetheless.

I met with the crew again and conducted a biomechanical analysis to determine the potential reduction of musculoskeletal disorder risk if a hypothetical trolley were used. The crew contributed to this process. They clarified or corrected the assumptions that I made so that I could consider aspects like the exposure levels, movement patterns, tool use, and weights. I watched them do the task, and they simulated scenarios for me, like when someone lost their grip during tyre manoeuvring. I took images and recorded videos. I measured the reaches and examined the equipment.

I was advised to steer clear of the forklift because of the inherent dangers with the risk of pedestrian contact or collision, and I was provided space to work outside the restricted areas.

The workers were determined to change this task, yet they whispered if line management were nearby. Something shifted, and this task change became their statement of rebellion. The risk assessment was conducted with ergonomic software[1] to consider the instrumental functional movement and biomechanical work demands, such as exertion, exposure, posture, and repetition of movement elements. Elements of pinch points, cognitive load, and work pace/stress were also considered. The software was chosen because it fulfilled the tenets of a risk assessment tool that was appropriate for manual task injury prevention (Burgess-Limerick, 2003), including the following:

- Suitability to a complete range of manual tasks.
- Capability to synthesise the assessment of a range of interacting factors that are biomechanical, cognitive, and psychosocial in nature.
- Unique assessment of injury risk to all associated but independent body regions.
- Comprehensive assessment of risk that incorporates the guidance about threshold tolerances without false levels of presumed precision (because of the significant variability among humans, tissue capacity, and work performance).
- Capable of determining risk factors that lead to acute and cumulative musculoskeletal disorders
- Determination of the level of the relative severity of independent risk factors per task exposure to inform design concepts and control strategies.
- Suitability for use by generalist safety and health teams should re-assessment be required to contrast any changes to the design of the task following baseline investigations (Burgess-Limerick, 2003).

The biomechanical risk ratings showed an extreme risk to the lower back and arms and a high risk to the shoulders and legs, mainly owing to exertion and awkwardness of postures. Also, there was the potential for cognitive distractions and perceived time constraints, which can contribute to workload exertion and risk for musculoskeletal disorders (Macdonald & Oakman, 2015). There were high cumulative risks to the lower back and arms and moderate cumulative risks to the shoulders and legs, owing to the combination of exposure levels, movement, exertion, and repetition. The task occurred in an open-air, sheltered workshop, so weather conditions could also factor into the risk considerations. This workshop was in the subtropics, so heat and humidity could add to fatigue. However, winter was cool to cold, which can add to reduced circulation, either circumstance a factor in musculoskeletal disorder risks (Burgess-Limerick, 2003).

If a suitable trolley were used, and the forklift was no longer needed, these risks would be significantly reduced by 80% in the back and arms for acute risk reduction

---

[1] https://www.ergoanalyst.com/

and more than 69% for the arms and back for cumulative risks. Also, productivity would improve because of the re-assignment of the forklift operator to other tasks and the expedience at which a wheel could be removed and manoeuvred within the workshop, an estimated savings of two-person hours each time the task was done (removing and replacing the wheel). This information was reported and reviewed through appropriate channels, including the regional and state management, and the trolley trial was approved.

REALISATION PHASE

The workers procured a trolley for trial, per the advice from the paver manufacturer (Figure 6.2). They trialled it and loved it. The crew told me about their satisfaction. They met their design objectives: the new trolley saved time, added efficiency, and moved them out of harm's way of mobile plant and from heavy lifting. Leveraging this success, another workshop within another region of the state purchased the same trolley. The project was celebrated among our crew, yet the team leader was affected, and so was his team. He started searching for a new job. He had been blamed for being somewhat of a "rebel-rouser" because of his passion for this project and his voice when speaking to senior management, above the line management, who initially refused his efforts. The others were sullen and uncertain of their future while working with the manager, who was in line for a promotion. The project was lauded by the state general manager, shared in an industry conference paper, and celebrated by way of communication in a national company newsletter. Notably, the line and capital equipment managers were credited with the project's success, and it was their photos displayed in the newsletter, taken with the new paver wheel trolley.

FIGURE 6.2   A wheel trolley that eliminates forklift use.

The project was successful if evaluated by its design objectives to reduce the risk of fatality or injury and add to productive, efficient work solutions for quality improvement. However, if measured by the effect on team morale, it had gone south. Post-rebellion satisfaction was short-lived because the reality of their job design and reporting lines meant that the workers felt less enthusiastic about making a positive change again. They were demoralised. The egos and perceived threats by management were discomfiting, and they wondered how they could be inspired to innovate or adapt tasks to improve design again. This is what I learned:

- Ergonomics workshops and training can inspire workers to consider task re-design, either for new tasks or those that had been reported as hazardous but never actioned. The training can invigorate workers to act.
- Shared problem-solving, like that which can occur through task re-design, is a hallmark of adult learning.
- Manual tasks represent the fabric of daily life for field workers; taking an interest in what they do and how they work is essential to their engagement.
- Consulting workers is in line with legislation in several countries, and participative ergonomics projects discharge the obligation of the duty holder (WHS Act, 2011). However, the involvement of workers should be sought in genuine and supportive ways, not just to meet compliance standards.
- Line management might resist change, even though there may be support from those above them and the field staff. Involving the line management as change agents is vital to success. Somehow, this must be done.
- Ego is dangerous if it impedes changes to safety and productivity.
- As an advisor, one's efforts are restricted to influence rather than command. It is difficult to be supportive when observing head-shaking poor management practices. It would be interesting to know what business would look like if work design strategists, with human factors and ergonomics training and ethos, were elevated as decision-makers with more power to effect change.

STUDENT-CENTRED CURRICULUM DESIGN

I am fortunate to consolidate and extend my learning in human factors and ergonomics through ongoing teaching and research activities with several universities. Dipping my toes in the proverbial waters of education helps me apply a framework to my consultancy that is supported by evidence. It continually challenges me to extend my thinking or, when teaching, to explain the art and science of human-centred, good work design in digestible terms so that it is better understood. In this case, the benefits of my professional work were applied inward. That is, I had the chance to work with a colleague[2] and use human factors approaches to address the curriculum design

---

[2] I extend my gratitude to my colleague, Dr Anita Hamilton, with whom I enjoy my work and from whom I have learned a great deal. I acknowledge Dr Hamilton's contribution to this course re-design and the journal paper that we wrote to describe the experience in academic literature.

of an undergraduate occupational therapy course. This meant that a student-centred process of the course design was embraced at a university level of education.

Luckily, I worked with a colleague who was welcoming of change and new technologies. Continual improvement is part of her ethos. We co-taught this university course and often shared ideas about teaching innovations or content specialties. This was a unique course: students were asked to learn about psychometric tools for standardised assessment, pretend that they were a client newly diagnosed with multiple sclerosis, and, conversely, act as a therapist evaluating their class-partner when they role-played their experience with multiple sclerosis (Figure 6.3). The students assessed different performance aspects weekly, presented their findings in mock case conferences, and presented formally to their professional peers when describing the assessment tool that they were selected to study in detail. There were weekly quizzes, and the course was offered as a "flipped-classroom", with online learnings and content engagement expected of students before the in-session facilitated tutorials (Cobb, 2016).

While these educational aspects are well in line with the professional role expectations of an occupational therapist, they were new to these second-year undergraduate students. When students struggled with the elements of clinical assessment, role play, or formal presentations, we wondered what could be done to elevate their learnings and shift anxiety into constructive learning. There were coaching tips that we could provide during the course delivery. Still, once a university curriculum is approved and implemented, a formal process must be undertaken to redesign the course. A new outline must receive central approval, ensuring that the changes meet accreditation standards and guidelines as a measure of quality control. We committed to

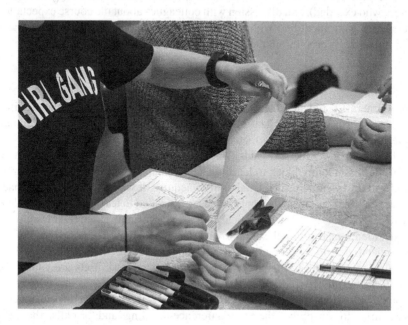

FIGURE 6.3 Occupational therapy students as mock clients and mock therapists.

reviewing the course formally for the next offering and learning from our teaching and the student experience.

DISCOVERY PHASE

A human factors approach to competency (Zupanc et al., 2015) and curriculum development (Pazell & Hamilton, 2020) was undertaken. We decided to learn from the student experience and invited past student involvement in the course re-design considerations. While these students were entering their third year of study, a busy term peppered by fieldwork experiences, we managed to attract a focus group of six. These students acted as intermediaries to explain our conversations and represent the needs of more than 40 of their student–colleagues who otherwise had difficulty juggling their schedules to attend our meetings. Since occupational therapy is a course grounded in client-centred construction of health and systems theories of person–place–activity–performance (Baum, Christiansen, & Bass, 2015), involving students to co-create improved educational models was welcomed by my colleague and the students.

Compatible with this ideology, we dissected the students' roles in this course. We identified six student roles: a generalist student, a mock client, a mock therapist, an evaluator of standardised assessment tools, a clinical documenter, and a case conference presenter. Each of these roles was dissected into nine cognitive components to frame the requirements: knowledge, skills, abilities, tactics, decision-making, situation awareness, heuristics, interpersonal skills, and intrapersonal skills (Pazell & Hamilton, 2020). Through course curriculum review; semi-structured and informal interviews and focus group meetings with students; review of assessment performance, especially the outliers (i.e. the work of past students who struggled, and that of those who excelled); and discussion with colleagues about the course expectations, 91 elements were identified within these components. That is, 91 aspects that we could influence or prop through course design and teaching leadership.

DESIGN PHASE

In the "generalist student" role, a student was expected to exhibit competence in the computer interface with the learning management system to engage with content before tutorials. We elected to focus on providing more early-stage assistance and Web-based video instructional aids. We did not want the computer interface acting as a primary stumbling block to learning key content. While computer literacy was considered an essential professional skill to acquire, it was not one that they had to achieve solo. In their "mock therapist" role, they needed to learn how to create defensible daily and summary-based clinical case notes. We emphasised their training and gave detailed scenario-based examples and tutorial content to develop these weekly. The student-advisors suggested less formal assessment of case conferences. While we struggled with letting this go, since case management skillsets are vital professionally, we negotiated trade-offs, like converting case conferences into weekly tutorial practice sessions without formal assessment. We worked with our colleagues in other courses to consolidate the case conference teachings and formalise the assessment in subsequent years.

Another interesting finding was the students' difficulty in conducting home and workplace assessments, constrained by time and the lack of understanding about what clinical symptoms to profile in different environments when acting as a mock client. Two changes arose. First, we converted the partner-based home assessment requirement to a simulated immersive environment using 360° camera technology that could be viewed either as a video or as an interactive experience using a headset. My colleague and I recorded four home-based and four work-based experiences so that they were standardised, and we knew what to expect and how to coach students. We punctuated these visual experiences with some narration and notation to explain environmental aspects, like trip hazards or narrow spaces to navigate in the home. This was designed to help the students in client and therapist roles to prompt their exhibition of symptoms or assessment focus. Second, we helped the mock client by creating a symptom list on small mixed and matched cards. In each tutorial, a student could select some random symptoms to help them formulate their presentation, which became the challenge for their mock therapist student-partner: to determine what symptoms were on display and how this affected the mock client's functional presentation.

Their contribution to the curriculum design impassioned the student-advisors. So much so, they agreed to be videoed to provide their "pearls of wisdom" when asked by my colleague to do so. These videos were played during tutorial sessions and acted as ongoing peer-based coaching support for students when the course ran in the subsequent term. A couple of the student-advisors agreed to come into the tutorial sessions to spend some solo time with the new students. They provided candid insights and advice about how to get through the course.

REALISATION PHASE

I learned a lot during this course re-design. No matter the wisdom or fresh insights I believe that I can lend to a situation, I am one voice. The role of the human factors professional is exceptional when allowed to be a catalyst for change, and when subject matter experts, those most affected by the design change, partake in the process as design partners Notwithstanding, in the workplace, these methods discharge the obligation of the duty holder and show compliance with workplace legislation in Australia and elsewhere (WHS Act, 2011). Good course design can boost student satisfaction (Lee, 2014). Also, accreditation standards exist in some programs, like in Occupational Therapy (OT Council of Australia, 2018), which suggest that "students have opportunities to be represented within the deliberative and decision-making processes of the program" (p. 8). This was supportive of the co-design methods that were undertaken.

Innovation is possible if the milieu is right, unfettered by management who say "no" to the methods, and inspired by colleagues who share the vision of co-constructing new ways of working. Friendships were formed, and others consolidated during this course redesign process, including ongoing positive relations with the student-advisors who are now graduates and working professionals. With ego set aside, change was made possible through an uncomplicated process, and positive feedback was associated with the adaptations. This made it a successful case. Tackling an entire university programme in such a way would be novel and, likely, harder to sway given the influence required of more decision-makers, but, as with most ergonomic changes, small wins pave the way for instrumental change and system re-design. Some of the students involved as advisors or participants in the new programme could very well be those who direct more of the system-wide revolutions in the design of the work that awaits them. It shall soon befall them to advance winning approaches representative of a human-centred organisation (ISO 27500:2016).

REFERENCES

Work Health and Safety Act. (Austl). (2011). https://www.legislation.gov.au/Details/C2018C00293

Baum, C. M., Christiansen, C. H., & Bass, J. D. (2015). The Person-Environment-Occupation-Performance (PEOP) model. In C. H. Christiansen, C. M. Baum, & J. D. Bass (Eds.), *Occupational therapy: Performance, participation, and well-being* (4th ed., pp. 49–56). Thorofare, NJ: SLACK Incorporated.

Burgess-Limerick, R. (2003). Issues associated with force and weight limits and associated threshold limit values in the physical handling work environment: Issues paper commissioned by NOHSC for the review of the National Standard and Code of Practice on Manual Handling and Associated Documents. http://burgess-limerick.com/download/d2.pdf

Burgess-Limerick, R. (2018). Participatory ergonomics: Evidence and implementation lessons. *Applied Ergonomics, 68*, 289–293. DOI: 10.1016/j.apergo.2017.12.009

Bushe, G. R., & Kassam, A. F. (2005). When is appreciative inquiry transformational? A meta-case analysis. *The Journal of Applied Behavioural Science, 41*(2), 161–181. DOI: 10.1177/0021886304270337

Cobb, W. N. W. (2016). Turning the classroom upside down: Experimenting with the flipped classroom in American government. *Journal of Political Science Education, 12*(1), 1–14. DOI: 10.1080/15512169.2015.1063437

ISO. (2016). ISO Standard 27500:2016: Human-Centred Organisations. International Standards Organisation.

Lee, J. (2014). An exploratory study of effective online learning: Assessing satisfaction levels of graduate students of mathematics education associated with human and design factors of an online course. *The International Review of Research in Open and Distance Learning, 15*(1), 111–132. DOI: 10.19173/irrodl.v15i1.1638

Macdonald, W., & Oakman, J. (2015). Requirements for more effective prevention of work-related musculoskeletal disorders. *BMC Musculoskeletal Disorders, 16*, 293. DOI: 10.1186/s12891-015-0750-8

Occupational Therapy Council of Australia Ltd (the OTC). (Dec 2018). Accreditation standards for Australian entry-level occupational therapy education programs. The OTC. https://www.occupationaltherapyboard.gov.au/Accreditation.aspx

Pazell, S. (2018). Good work design: Strategies to embed human-centred design in organisations. [Doctoral dissertation: University of Queensland]. Sustainable Minerals Institute. https://espace.library.uq.edu.au/view/UQ:3e5556a

Pazell, S. (2021). Design for workplace diversity: A human-centred approach. ViVA health at work. https://vivahealthgroup.com.au/ergonomics-resources/

Pazell, S., & Hamilton, H. (2020). A student-centred approach to undergraduate course design in occupational therapy. *Higher Education Research & Development, 40*(7), 1497–1514. DOI: 10.1080/07294360.2020.1818697

Zupanc, C. M., Burgess-Limerick, R., Hill, A., Riek, S., Wallis, G. M., Plooy, A. M., Horswill, M. S., Watson, M. O., & Hewett, D. G. (2015). A competency framework for colonoscopy training derived from cognitive task analysis techniques and expert review. *BMC Medical Education, 15*(216), 1–11. DOI: 10.1186/s12909-015-0494-z

7 Reshaping Lifestyle Changes in a Heavy Weight World

Keith Johnson
Fulton Hogan

CONTENTS

Several years ago, I worked at an open cast coal mine as a safety and compliance superintendent. The average age of the workforce was between the mid-30s and 40s with very minimal turnover. The mine was close to a local town where most of the miners lived. The nature of mining is a sedentary job generally. The bulk of mining operations relates to operating a mobile plant (e.g. excavators, bulldozers, trucks) whilst also being seated for the task. This, in combination with a lack of activities outside of work due to long working hours at the mine and general poor health due to genetics and lifestyle choices (Suckling, 2017), led to a proportion of the workforce being overweight or obese.

As such, part of the workforce became too heavy for some of the truck seats in a specific type of mine haul truck. The standard seat of the truck had a weight-tolerance rating of 120 kg. The workforce complained that the trucks' seats would bottom out when going over bumps on the haul road, even with the gas struts in the seat set at maximum level. Several incidents were reported on site relating to seats bottoming out, with initial causation thought to be related to the poor design or maintenance of seats. However, postincident reviews (e.g. vehicle seat servicing history, seat functionality, road surface compliance conditions, and review of operator training

DOI: 10.1201/9781003349976-7

records) suggested that the seats and systems were compliant. Furthermore, the analysis of demographic data from the incidents suggested that the latter mainly involved overweight or obese workers.

Back then, we contemplated that there were only two alternatives to dealing with this problem. The first option, from a system design perspective, would be to replace all 120-kg weight-tolerance rated seats with 200-kg ones across the whole fleet to accommodate the overweight workers. However, the cost to swap out the seats would be quite high. The prices for the seat itself started from $3,000.00 per item, depending on country of purchase with exchange rate and subsequent freight costs. This cost excluded the necessary labour costs for removal of the old seat and installation of the new seat, the downtime of the machine and potentially not being able to sell the old seats and having to store them. From a reasonably practicable perspective, the company initially contemplated that the seat replacement cost was 'grossly disproportionate to the risk' (Safe Work Australia, 2011).

The second alternative was to terminate the overweight workers' employment because they could not meet the inherent requirements of the job. This option was quickly dismissed. Apart from ethical reasons related to first exploring other alternatives, the workforce came from a close-knit local town and the site was heavily unionised. Hence, taking that stance may have ended in strike movement by the union. The latter would subsequently incur costs arising from potential wrongful dismissal claims and/or breaches of industrial relations legislation of Enterprise Bargaining Agreements for the employer. This could be coupled with reputational risk of community backlash against the company.

Indeed, there were also weight loss and general health initiatives we could consider. However, these were not thoroughly researched or challenged as an alternative due to the combination of the following factors:

- My general lack of understanding of the science behind such initiatives.
- Lack of openly accessible research on the outcomes of health initiatives.
- The cost to implement such an initiative. The employer should allocate funds to secure support for workers by nutritionists, medical doctors, dieticians, and/or gym memberships. This kind of support is quite costly and would generally occur offsite due to the lack of in-house company resources. Over time, and with regular appointments per employee, such an initiative would lead to thousands of dollars expenditure per employee.
- The time lag between the implementation of health initiatives and actual results for the overweight workers. This would have been several months, and because the overall success of such an initiative was not guaranteed, it could have been an overwhelming failure.

After general consultation with the workers and management team and some heavy influence from union representatives, it was determined to proceed with the first option above but only for the newest fleet. The reasoning was that the remainder of the fleet was ageing and close to its end-of-lifecycle, and the seat replacement was cost prohibitive. On the other hand, the newer fleet had a much longer life expectancy, which made the purchase commercially viable to accommodate those in the heavier weight range. Thus, the replacement of the 120-kg weight-tolerance rated seats with ones having a safety weight-tolerance rating of 200 kg would expectedly fix the problem of our overweight workers bottoming out on truck seats. This decision made sense at the time, because it seemed like a good engineering solution and aligned with the concept of the well-known hierarchy of risk controls.

Hence, the site had managed the risk and would not ask workers to operate any machinery without enjoying adequate safety levels. At the same time, the decision was the path of least resistance as it was also supported by the mine's workforce and union interventions. Moreover, this option had been successfully employed at an associated mine site, so it was deemed to be aligned with 'standard practice'. Surprisingly, though, whilst the replacement of seats fixed the immediate design-related issue of seats bottoming out and subsequent injury risk to the truck operators, it led to more complicated issues.

First, the workers didn't seem to implement anything at a personal level to reduce or manage their weight. Either their weight remained the same, or they became heavier because of their sedentary work coupled with long shifts and limited access to and knowledge about healthier meal options. Furthermore, the workers did not have any work-related incentives to lose weight, especially after the new seats accommodated heavier operators. Based on anecdotal workplace conversations, the workers felt there was no reason to abstain from the consumption of unnecessary calories. Second, the workers that exceeded the 120 kg weight rating could now only drive the trucks with the 200 kg seats. Since the trucks with the newer seats were the most recent in the fleet and had better cab operating facilities, this created a logistical challenge and inequity among fleet allocation. Workers under 120 kg felt marginalised from the newest fleet by workers that could only drive the trucks with 200 kg seats because of the company's new rules and, in hindsight, lack of distributive justice. Simply put, lighter workers could only drive trucks that had seats with a maximum rating of 120 kg, those trucks being the older ones in the fleet.

Third, the mine rescue team encountered further complexities. If a worker had a medical issue (e.g. a heart attack) or another emergency in/on a truck with 200 kg seats, trying to extricate the worker safely without any injury to the worker or the mine rescue team became complicated and risky. Moving an extra heavy worker from a cab that was a minimum of 4 m from ground level created manual handling risks for the rescue team, and health and safety risks in general. The exertion arising from heavy and uneven (human) loads could result in musculoskeletal disorders, and there was a risk of falls whilst carrying a person from a high cab to the ground. Fourth, as workers exceeding 120 kg could only drive one type of truck, the scope of the mine to cross-train these workers in other machines and activities (e.g. dozer or grader operator) was limited because this other type of machinery generally had a seated weight rating of only 120 kg.

Consequently, whilst the uprated seats fixed the issue from an engineering perspective, the measure had no impact on workers' health and generated unintentional side effects. Thus, with the benefit of hindsight, replacing the seats was a poor call and should not have been undertaken. It seemed a good idea at the time, but in long term was counterproductive for the workforce and the company and set a poor standard for future risk management strategies and as a general benchmark for best practice in the wider industry. On reflection, I attribute all the above to the fact that the uprating of the seats was not based on a systematic and collective risk or change management approach and a well-thought plan. The decision was mainly the result of influences from external stakeholders, ignorance by a poorly informed safety department, including myself, and the pressures on the employer to act the soonest possible.

THE JOURNEY TO SUCCESS

After realising the complexity of safe seating weight tolerances, we introduced remedies to better manage overweight workers and the situation overall through better system design per our understanding of the concept of total worker health, as presented in more detail below. The other side of success regards my professional journey since then. It is now based on research, study, and real-world experiences during which I have been able to develop, implement, and make better-informed decisions. The part elaborating this journey includes the various steps involved in better supporting overweight workers, starting from the recruitment and onboarding systems and processes, including a Health Management Plan, continuing with performance management, workplace initiatives, anthromechanics in equipment design, and concluding with the legal position about where/when an overweight worker may be dismissed from the business.

REMEDIES AT THE SPECIFIC MINE SITE

The first remedy adopted to ensure improvement for both the workers and the employer was that the mining trucks with uprated seats remained on site for some time and were later sent to a mine site in another country. No other seats on the site were uprated beyond the manufacturer's specification on the original seat. This demonstrated equality across all workers because all seat weight ranges remained the same across the fleet. More importantly, it set a minimum standard or a 'one size fits all' approach where design changes would not be made to accommodate those who could not meet the minimum standard. Indeed, this does not reflect equity where the work system serves the needs and characteristics of diverse workers and supports those most marginalised. However, in conjunction with the other measures and initiatives explained below and from a pragmatic perspective, equality or equal access to the seating types in truck fleets was deemed the most feasible approach back then.

Second, in consultation with the workforce, procedures were also developed and discussed about what the new company standard would be for body weights and access to trucks and other equipment. Any worker that remained excessively heavy was provided with new strategies and requirements. If a worker's weight was above the seat rating, the worker was stood down from that substantive role and deployed to another area of the business, on some form of suitable duties, which was agreed between a medical provider, the employer, and the employee. Admittedly, the suitable duties roles were difficult to effect because, generally, the tasks were not meaningful, or inspiring, or had a short working window and included tasks such as photocopying, filing, or in-house training. However, this was only a first temporary step of the whole support programme. Additionally, the worker was placed onto a Health Management Plan, provided with five free consultations with a nutritionist, and given subsidised access to the local gym.

Furthermore, during the recruitment process, candidates were advised of the weight ranges of the equipment regarding seat weight ranges for all plant and equipment. This formed part of the pre-employment medical screening. Also, induction training included better explanation of weight restrictions for equipment whereby the pragmatic design specifications were explained so that workers understood the escalation of their risks of developing musculoskeletal disorders and exposure to whole body vibrations if their weight exceeded the seating design tolerances. In parallel, the company launched a healthy eating campaign to raise workers' awareness. To support this, the workers were provided with subsidised health subscriptions, fruit platters, and other healthy refreshments which were placed into the lunchrooms. As an employer, we wanted to provide environmental conditions that supported good nutritional choices (Cohen and Farley, 2008).

The outcome was that six overweight workers got under the maximum weight over the course of two to six months. Moreover, due to the matter being a topical issue and a regular subject of conversation in the workplace, awareness was also heightened.

THE AUTHOR'S JOURNEY

Recruitment and Onboarding

The first key aspect I have pursued in the management of overweight workers is to identify them during the recruitment process and before they enter the company. Undertaking a robust pre-employment medical examination is critical and can be as simple as a health questionnaire that identifies a candidate's health and eating habits and weighing the applicant on scales to calculate their body mass index (BMI). This index helps to determine a person's healthy weight range for his/her height (The Heart Foundation, 2018). Depending on the person's weight, the identification that the job candidate could be a risk to the company can be and has been a catalyst in offering or not employment (e.g. concerns about climbing stairs, bottoming out seats on equipment on rough roads, or inability to manoeuvre in confined spaces).

A precondition to the pre-employment medical assessment is the employer having a solid understanding of the equipment used in the business that has a weight safety rating. This can include items such as weight ratings for seats based on the recommendations of manufacturers (e.g. office chairs, and vehicle seats), portable and mobile ladders, and safety harnesses for working at height. The company maintains an updated list of all equipment with their weight limits that we provide to the pre-employment medical provider. This way, the latter becomes aware of the suitable weight range for specific jobs.

Systems and Processes

I have found it critically important that the company should also have robust systems in place in the form of policies and procedures for the minimum standard and expectations for workers' health related to their weight during their employment. Workers should also be supplied with a job description that outlines their legal obligations and duties in the workplace to take reasonable care for their health and safety to the extent this is possible and under their control. Another element around policies and procedures is the frontloading and education of frontline staff. The consideration of overweight workers might not necessarily be at the front of the mind for frontline leaders. However, those persons should know specific elements, including definitions of the terms 'overweight' and 'obesity', implications of having overweight and obese workers in the workplace, management of overweight and obese workers, and subsequent legal obligations and pre-employment considerations.

Workplace Initiatives

There are varying workplace initiatives to assist workers to maintain good eating practices or access resources that encourage healthy eating habits. One possible problem is when managers do not have a comprehensive understanding of how work can impact a workers' diet. This can be as simple as understanding the working environment. According to Pro Choice Safety Gear (2016), diet issues for workers relate to time pressures to get the job done; constantly changing worksites, as it typically happens in the construction industry; or limited availability of food due to a lack of lunchrooms, refrigerators, etc. Because workers can be hungry, tired, and time-poor, it is easy to default to vending machines and food vans, because it is 'food to go'.

In recent years, I have sought to influence good eating practices on-site through offering nutritional food and fruit bowls; providing access to lunchrooms/crib huts with microwaves and refrigerators to assist with pre-prepared healthier meals; providing skim milk instead of full cream milk and artificial sweeteners instead of sugar; and using a food environment audit tool (Griffith University, 2016) to better determine the eating habits of the worksite. Improving food choices and amenities in the work environment can lead to healthy eating becoming an extension of the work organisation (Lavallière et al., 2012)!

Furthermore, when looking for a minimum standard for implementing a weight loss initiative programme, I have used the baseline assessment based on the 'Healthy Workplace Framework and Model' (World Health Organisation, 2010). This framework covers eight steps from mobilising and assessment through to planning and acting, and then reviews and improves. It can be used in any health initiative roll-out but is instrumental in the foundation requirements of Plan-Do-Check-Act for a successful weight loss programme.

For an in-house weight loss programme, I have implemented the Mediterranean weight loss programme (Martin et al., 2019). This is a 12-week programme to assist in reducing body fat, and it is prefaced with a food frequency health questionnaire and information sessions. The programme includes promotional awareness about healthy eating and personal obesity risks, and it focuses on healthy foods such as high intakes of fruits, legumes, whole grains, nuts and fish, and low intakes of dairy, red meats, and alcohol. However, when I implemented this programme, it was only effective for those that stayed for the whole 12-week course. Unfortunately, such programmes can have a high dropout rate due to several reasons such as a lack of motivation, the initiative not being suitable for the worker or the worker not being able to adapt to the suggested lifestyle change.

Other 'blanket' weight loss and healthy eating initiatives I have implemented or managed over time with reasonable success include the following:

- Healthy snacks and drinks in vending machines. Whilst the traditional options are still available, healthy snacks provide variety and more options.
- On-site health assessments that include the assessment of weight and BMI on a voluntary participation basis.
- Training sessions conducted by nutritionists and dieticians that speak about (un)healthy eating and practices that can frontload the workers with scientific and research-based information.
- Nutritional information posters in lunchrooms/cafeterias.
- On-site exercise facilities. This is more applicable to larger projects and remote sites with camp accommodation.
- Discounts or waived fees for gym memberships.
- Diabetes prevention programmes.

Health Management Plan

When a worker has been identified as being overweight and not meeting the inherent requirements of their role, in consultation with human resources/employee relations staff, my process has been to hold an initial conversation with the worker to discuss my concerns and offer the opportunity to the employee to respond. Where a mutual agreement is met and the worker has agreed to lose weight, he/she is then placed onto a Health Management Plan (Leighton Mining, 2009).

The Health Management Plan includes employee and employer information (e.g. names and contact details) and continues with the Action Plan. The latter is the key part of the plan as it relates to the weight goals agreed (e.g. the target weight to be lost by specific date/s) as well as the support services required. Those services might include access to a dietician and/or a nutritionist to assist the worker with a menu plan and fitness goals and access to fitness facilities such as a gym or personal trainer if that's what makes sense to the employee. What I have experienced is that whilst the company may be paying for the nutritionist or dietician, due to privacy reasons, the nutritionist or dietician may not share information/reports with the employer. Therefore, such information needs to be obtained from the worker, where possible.

All these added services generally come with a cost; hence, this needs to be a consideration for the company as well. I have had examples where the worker has failed to achieve the necessary weight loss whilst accessing such services, meaning that partly those cases could be seen as wasted effort and investment. Where this occurs, employees must have regular catch ups with their nominated nutritionist/dietician to continually monitor and review their progress, their barriers, and whether they remain aligned to the plan, and then determine the best method to get back on track and towards a successful outcome. The final component of the Health Management Plan is the approvals section where the employee and other stakeholders and persons involved (e.g. manager, medical professional, injury manager) sign the documents. This means the plan has been read by each stakeholder and is agreed upon and endorsed.

Nevertheless, it is important to note that the initial conversation can be quite daunting for the worker because they are generally told they cannot undertake their normal role anymore (e.g. truck or grader operator) and must assume another role. The latter can be something they do not want to do and could be as menial as filing or photocopying. The worker can also become wary about having their employment terminated, and the employee can ask for clarity on this topic. The rest of the workforce can also see what is occurring with the specific worker as he/she is absent from their normal place of work. Thus, this can add to the worker's humiliation of being stood down from their normal role. It is just as important to consider the worker's mental health during the Health Management Plan process, which is generally coupled with the company's mental health and well-being programme and Employee Assistance Program. The latter is available to the employee as a free counselling service provided by the employer and is confidential.

Anthromechanics in Equipment Design

Another matter for consideration with aiding overweight workers is through equipment design that the worker uses and can be reviewed through the research of anthromechanics. Kumar (2008) suggests that anthromechanics is the connection between anthropometrics (i.e. measurements and capabilities of the human body) and mechanics, with the research investigating the concerns of mechanical design and fit as it relates to anthropometrics. Interior car design is an example of anthromechanics, whereby the vehicle's interior is designed to meet the needs of the greater population and not a limited quota, and the ergonomic goal is to design a product or system comparable to the needs and limitations of the end user (Kroemer, 2007).

Vehicle manufacturers design the interiors of their vehicles (e.g. foot pedals, armrests) to accommodate anthromechancial concerns (Bansal et al., 2009). Some car manufacturers have developed an 'aging suit' so designers have an understanding of the impacts of aging when operating a vehicle (Pultarova, 2016). The aging suit is worn over the body (like an outer metallic skeletal suit) and mimics health concerns linked to aging such as vision impairment and loss of movement range or flexibility. A similar suit could be developed that mimics overweight users so that car manufacturers could consider these anthropometrics in car design (e.g. the positions of the foot pedals or steering columns that are adjustable for volume).

Another example of anthromechanics is designing seat belts for overweight people. Research conducted by the University of Virginia (2019) suggests that thicker fat around the belly and waist inhibits a seat belt suitably grasping the pelvic region whereas a suitably engaged seat belt for thinner persons sits lower below the belly creating more resistance on impact. With this knowledge, researchers are trying to develop a new seat belt that factors in the weight of the occupants for the provision of timing and force level when a seat belt is engaged on impact. The research is also considering the application of inflatable seat belts where the belt inflation bag quickly expands across the body to disperse the crash forces and reduce the impact. The design ideas are all good examples of where manufacturers, designers, and companies should be considering the anthromechanical concerns as they relate to the overweight/obese in the workplace.

Performance Management

Another method that could be considered is using the best practice guide for managing performance (Fair Work Ombudsman, 2021). I have utilised this in the early stages of identifying and managing overweight workers who cannot fulfil their role or fail to comply with workplace policies and procedures. The best practice guide allows the employer and employee to develop a framework for goal setting. It also comprises an opportunity for constructive feedback and sets a benchmark for regular follow-ups about performance. I have utilised the performance system outline below to manage the issue of obese and overweight employees. Specific elements include the following:

- Expectations. Outline specific expectations during the employee onboarding and company induction with references to healthy weight range for employees for the safe use of plant and equipment. Expectations must also be clearly articulated in the Health Management Plan, with specific goals to be met, and acknowledged and understood by the employee.
- Template Agreements. This refers to the creation of the Health Management Plan as described above.
- Discussions/Feedback. This includes meetings with the employee to review their progress and any future needs or determinations.
- Reward. This is about recognising and rewarding the employee for achieving goals in terms of weight loss or when he/she has reached the final goal and is no longer on the Health Management Plan.
- Review. This includes conducting performance reviews over the course of the Health Management Plan and when the employee has not achieved their goal by the agreed deadline. In the latter case, action commences to manage the worker either through an appointment to another role (if appropriate and possible), or initiating the dismissal process.

Legal Position

The final element in the management of overweight workers is activated when all other forms of management and initiatives have derailed, all avenues have been exhausted, and the worker has not lost the required weight. It is at this time that the company needs to determine its legal position from the employee perspective that the worker can no longer (safely) do the job they were employed to do. Also, the company might now be keeping a worker in a role they were not employed for in the first place. If the worker cannot be re-deployed to another role and he/she has become a liability, the employer may then commence the process of terminating their employment through the relevant employment legislation and/or Enterprise Bargaining Agreement or similar.

In addition to the preceding management elements that must have been effectively adopted (e.g. work system design, Health Management Plan, performance management, and other health initiatives), a direction that a company can take and has been relied upon in matters I have had to manage is utilising legal precedent (a.k.a. case law) or other industry standard practice to justify and validate the company's position.

Importantly, in the early stages of the worker's weight management, the employer should have reviewed their legal position with reference to relevant legislation. When reviewing an overweight worker and their exposure to risks in the workplace, the company may have legal exposure related to its health and safety obligations even whilst implementing the rest of the initiatives I described earlier.

For instance, Section 19 of the Work Health and Safety Act 2011 in Queensland, Australia, mentions the employer's *Primary duty of care*, whereby the employer needs to ensure, so far as is reasonably practicable, the health and safety of workers engaged, or caused to be engaged by the employer. If prepared to allow workers to be seated on plant or equipment that does not meet the weight rating for the operator's weight, then the company has not fulfilled its legal obligations.

Similarly, though, workers may have also breached their own obligations. Section 18 (Work Health and Safety Act 2011 (Qld)) relates to the *Duties of workers* whereby the worker must take reasonable care for his/her own health and safety. When the worker's weight exceeds the safe working limits of a seat on plant and equipment, he/she has failed to ensure his/her own health and safety subject to the worker being aware of the respective limitations. These legal elements are crucial in the recruiting and onboarding stages. If the worker enters the company and the employer has not done its homework or exercised due diligence to check the weight limits of plant and equipment and/or advised the workforce about these, it will be much more difficult to terminate an employee's employment.

When it comes to citing industry standards on managing overweight/obese workers, the first notable example in Australia which sets a quasi-minimum standard for other employers was the case of Metro Tasmania bus drivers in Tasmania (ABC News, 2012). In this case, bus drivers who weighed more than 130 kg were taken off the job and put on lighter duties by the employer. The drivers were given six months to lose weight, and they were also offered free medical consultation and gym access to assist them with their weight loss. The company acted according to its policy. The company had concerns for the employees' health and safety in case the bus bottomed out and the seat could break, thus causing an injury to the worker possibly attributed to the mismatch between personal weight and maximum allowed weight of the bus seat.

Whenever I must deal with overweight workers regarding exceedance of maximum safety capacities, I often print and hand out the article cited above. I also use the six-month time frame adopted by Metro Tasmania as the same benchmark for the employees of the company I work with to meet the required weight range. In terms of the desired weight to achieve, the weight the employee and management normally agree upon in consultation with the medical doctor is a few kilograms below the maximum safety rating; this affords the worker some wriggle room. Similarly, after the weight range is achieved, the worker and management commit to follow-up sessions of about 15 minutes twice a month for three months. These sessions are to verify that the worker maintains a safe working weight and any potential hurdles going forwards that we can support and/or for which we can prepare.

A similar case regarded overweight miners at specific Bowen Basin mines in Queensland in 2012, who were stood down from work until their weight was under 120 kg before being allowed to return to work (Courier Mail, 2012). The Department of Employment and Economic Development, which was the mining regulator at the time, was consulted by the newspaper. The department stated that whilst there was no mining safety legislation that could '*impose minimum or maximum weights for persons working in the mining industry*', the Queensland legislation does require all mines to ensure they have safe operations and manage risk for their workers. This included fitness to work within the requirements of the safety and health management system. The department's position aligned with the employer's primary duty of care mentioned above.

The final element that an employer can rely upon in managing an overweight worker is legal precedence in the form of case law as this sets the minimum legal standard for similar cases. A prominent case is that of *Ranui Parahi v Parmalat Australia Ltd* (2015) FWC 7191. In this case, the worker was a cool room operator with a requirement to operate a forklift truck as part of his role. The forklift had a seat rating of 175 kg, and the worker weighed 165 kg at the time. In June 2014, the worker was stood down from work until he was able to manage his medical issues (i.e. sleep apnoea which posed a risk to the operation of machinery) and resume duties. The worker had been assessed by a specialist occupational physician who deemed him unfit for work and recommended standing the worker down. Subsequently, the worker was placed onto a treatment plan, which included weight loss (Lehrer, 2016).

However, when the worker was reassessed in February 2015, his weight had increased to 175 kg. The independent physician determined the worker could no longer undertake manual handling tasks or use machinery such as operating a forklift. The worker was subsequently dismissed because his weight rendered him unable to safely perform the requirements of his role. The worker filed an unfair dismissal claim with the Fair Work Commission, and the proceedings determined that the employer had valid reasoning in terminating the employment as the worker was incapable of safely carrying out the inherent requirements of his role (Lehrer, 2016).

I believe the few examples above outline adequately an employer's position whether it is about placing someone onto a Health Management Plan such as in the Metro Tasmania case, standing down someone from work until they reach a set weight range as per the Courier Mail article in the mining sector, or justifying why a worker is dismissed because they can no longer complete their role due to weight gain as per *Ranui Parahi v Parmalat Australia Ltd* (2015) FWC 7191.

CONCLUSION

The topic of weight management in the workplace can be prickly in nature and is a complex subject with challenges for everyone involved, including the impacted employees, management, and other key stakeholders. When contemplating solutions for addressing weight management in the workplace, it is imperative that the employer considers the work system with the highest priority. This can refer to the design of systems and processes, anthromechanics in equipment design, recruitment and onboarding, and other workplace initiatives, as opposed to focusing on the employee from the outset.

From a pragmatic perspective, the ideal situation of adapting continually the systems to each worker may not be feasible. Employees come in all shapes and sizes.

Therefore, whilst consideration needs to be given to the maximum possible inclusive design of the workplace and organisational systems, this might not be always practicable. Also, as I explained in the 'failed' case in the beginning of my chapter, rushing to change without a holistic assessment can generate more problems. This, engaging professionals with adequate and proven skills and experience, like certified ergonomists, should be one of the options to consider.

Notably, employers are the ones who first carry the responsibility for deciding and acting proactively. They must understand the design parameters of job requirements, including specifications and limitations of safe equipment use. Such an analysis, which can be performed and supported by ergonomists and other specialists, is important to support the adaptation of the organisation to restrictions and constraints outside its control (e.g. given equipment designs) whilst, in parallel, continually searching for more design-inclusive alternatives. The goal must be to support and influence employees, not to control them.

Furthermore, besides technical parameters, the integration of health and well-being measures can promote worker health by using the workplace as an environment which enables better health choices and habits of which workers might not be aware. However, whilst initiatives like the above must be subject to regular consultations, reviews, and updates, we cannot exclude cases where they do not deliver for everyone.

REFERENCES

ABC News. (2012). *Overweight bus drivers paid to go to gym*, accessed on 6 September 2021, https://www.abc.net.au/news/2012-11-13/overweight-bus-drivers-to-be-taken-off-the-job/4368322

Bansal, V., Conroy, C., Lee, J., Schwartz, A., Tominaga, G., & Coimbra, R. (2009). Is bigger better? The effect of obesity on pelvic fractures after side impact motor vehicle crashes. *Journal of Trauma*, 67(4), 709–714.

Cohen, & Farley, T. A. (2008). Eating as an automatic behavior. *Preventing Chronic Disease*, 5(1), A23.

Courier Mail. (2012). *Weighty issue hits miners*, accessed on 10 September 2021, https://www.couriermail.com.au/news/queensland/mackay/weighty-issue-hits-miners/news-story/bf03eaf3f13488f19b03d441f4ce33e2

Fair Work Ombudsman. (2021). *Managing underperformance, Best practice guide*, accessed on 29 November 2021, https://www.fairwork.gov.au/sites/default/files/migration/711/managing-underperformance-best-practice-guide.pdf

Griffith University. (2016). *Food Environment & Health Audit Tool - Food in Construction Queensland Project*, accessed on 18 September 2021, https://www.griffith.edu.au/__data/assets/pdf_file/0025/92266/Food-Environment-and-Health-Audit-Tool.pdf

Kroemer, K. (2007). *Anthropometry and biomechanics: anthromechanics* (2nd ed.). Boca Raton, FL: CRC Press, Taylor & Francis Group.

Kumar, S. (2008). *Biomechanics in ergonomics* (2nd ed.). Boca Raton, FL: CRC Press, Taylor & Francis Group.

Lavallière, M., Handrigan, G. A., Teasdale, N., & Corbeil, P. (2012). Obesity, where is it driving us? *Journal of Transportation Safety & Security*, 4(2), 83–93. DOI: 10.1080/19439962.2011.646386

Lehrer, B. (2016). *Sacked for being obese? Yes and no*, accessed on 11 March 2022, https://diffuzehr.com.au/sacked-obese-yes-no/

Leighton Mining. (2009). *Health Management Plan*, retrieved from Cintellate database.

Martin, I., Rojo, S., Becerra, X., & Vilar, E. (2019). *Successful implementation of a Mediterranean weight loss program to prevent overweight and obesity in the workplace*, American College of Occupational and Environmental Medicine.

Pro Choice Safety Gear. (2016). *Managers contributing to poor diets among construction tradies*, accessed on 20 September 2021, https://prochoicesafetygear. com/ppe/blog/construction-whs/tradies-poor-diet-construction/

Pultarova, T. (2016). *Ageing suit helps Ford engineers think differently*, accessed on 2 April 2022, https://eandt.theiet.org/content/articles/2016/02/ageing-suit-helps-ford-engineers-think-differently/

Ranui Parahi v Parmalat Australia Ltd (2015) FWC 7191.

Safe Work Australia. (2011). *Interpretive guide line — Model work health and safety act the meaning of 'reasonably practicable'*, accessed on 20 November 2021, https://www.safe-workaustralia.gov.au/system/files/documents/1702/interpretive_guideline_-_reasonably_practicable.pdf

Suckling, L. (2017). *Why New Zealanders are still among the fattest people in the world*, accessed on 9 March 2022, https://www.stuff.co.nz/life-style/well-good/motivate-me/94695382/why-new-zealanders-are-still-among-the-fattest-people-in-the-world

The Heart Foundation. (2018). *What's your BMI?*, accessed on 21 November 2021, https://www.heartfoundation.org.au/bmi-calculator

University of Virginia. (2019). *Engineering researcher works to make auto seatbelts safer for obese people*, accessed on 3 April 2022, https://phys.org/news/2019-02-auto-seatbelts-safer-obese-people.html

Work Health and Safety Act 2011 (Qld). Accessed https://www.legislation.qld.gov.au/view/pdf/inforce/current/act-2011-018

World Health Organisation. (2010). *Healthy workplace framework and model - background and supporting literature and practice*, accessed on 19 September 2021, https://www.who.int/occupational_health/healthy_workplace_framework.pdf

8 Indian Farm Tractor Seat Design Assessment for Driver's Comfort

Bharati Jajoo
Body Dynamics

While driving, you check the rear-view mirror often and view the scene that you passed. You have already passed it, yet you keep looking, checking to gauge traffic and speeds of other road users, checking if any safety hazards are still there and so on; although you are in the front, you still look back. Checking the rear-view mirror while driving is such an important habit to master.

The opportunity to share insights for a practice-focused book felt like a similar exercise: to review and appreciate learnings. What experience teaches you is to do more of the same or not do it at all. God's greatest gift to mankind is the ability to learn, improvise, and do things in a certain way. While thinking about writing this chapter, I pondered my learnings and insights from ergonomic design. One of the assignments that stood out in my memory, even after a decade, was the opportunity to work on the reduction of low back pain for tractor drivers.

Agriculture is the primary source of livelihood for about 58% of India's population, and so the role of modern tractors has become increasingly significant. Tractors have improved the capabilities of farm work and production fivefold. Tractor producers recorded their highest ever sales in 2017–2018. Selling over seven lakh units, a number expressively higher than the 5.9 mark in 2016–2017, the 22% increase in tractor sales indicates that farmers are opening to the idea of investing in multipurpose tractors, with their penetration improved from 1-per-150 hectares to 1-per-30 hectares.[1]

A leading farm tractor manufacturer contacted me with a request to help them make the driver seat more comfortable for the operators of farm tractors. At the outset, it sounded very interesting. My instant joy of being able to work on tractor seat design was beyond boundaries. My father was an orange orchid farmer/grower, and my grandfather was known for the advanced agricultural practices of his times in our area. I grew up partially in an Indian village. As a child, I had many rides in a tractor that were considered a big advancement from the bullock cart.

At that time in India, transport in villages had fewer choices. Tractors were used from agricultural tasks to ferrying people to their field jobs, and from wedding

---

[1] https://kmwagri.wordpress.com/2018/09/19/importance-of-using-tractors-in-modern-agriculture-kmw-agri/

DOI: 10.1201/9781003349976-8

processions to carrying patients to the hospital in an emergency. I saw the utility and importance of these vehicles from many angles. Therefore, this opportunity to work on the improvement of the farm tractor seat design was exciting. Being a farmer's daughter, I could contribute my part with the right product design from a human factors and ergonomics (HFE) perspective. With the right HFE design, I could help to alleviate the most common and, at times, debilitating back injuries of farm tractor drivers.

I believe that the right system and work design leads to prevention, efficiency, and ease of operations for the users. Notably, musculoskeletal injuries happening at the field level during agricultural activities are often not recorded, meaning little to no data are collected. It is an unorganised sector in India and lacks regulations for employee health, safety, etc., although such protective provisions may be a norm in many western countries.

Traditionally, Indian tractors are driven by hired drivers rather than owners. However, a market survey conducted by the client revealed that customers were seeking increased comfort coupled with better product performance. The market research also pointed to customer feedback indicating complaints of low back pain. Although no precise statistical data on low back pain were available, the feedback was compelling enough to encourage the client to build a new prototype and, amongst other objectives, address the low back discomfort issues related to tractor seat design.

Admittedly, I felt nervous when I was assigned this project as I did not have experience in the ergonomic design for tractor drivers. I was unsure of how I would go about it. Nevertheless, with enthusiasm and a solid understanding of the basics of ergonomics from more than 15 years of experience, I embarked on this exploratory journey. During our initial communication, it sounded difficult to lay down the scope and expectations of the team in charge without sighting the product. It was about a prototype and did not exist yet! Hence, the request to assess the currently available product proved helpful.

One thing I wished at that time was to have a mentor or any senior to guide me on unusual topics or projects. They do say that you are alone at the top, and that is the feeling I had while we were in the process of defining the scope and deciding milestones. Consequently, my first visit to the client's head office to review tractors and hold discussions was a mixed bag of experiences. On the one hand, I felt excitement for doing something new to me, and, on the other hand, I was unsure of not knowing how I would go about it in terms of defining the scope, precise data collection methods, risk identification, etc. I was thankful that I had previously read about tractors and their basic mechanics. This meant that I could engage in somewhat relevant discussions and understand more about the mechanical engineering aspects of the product. However, during this first visit, I realised I was unaware of the various tractor controls and terminologies used. Everything was new, and the product functioning was so different from my previous design experiences in designing for low back support, chair seat design, etc.

Luckily, an engineering team member that gave me the information was enthusiastic about providing the details about the mechanics, the engine horsepower, how the engine worked, and what mechanical engineering improvements were in the planning process. "Madam", he said, "all you need to do is give us a good seat design and our prototype will be ready in a few days!" In this visit, I was trying to gather and filter information relevant to ergonomic risk factors identification. However, although I had been given an overload of information, there was nothing relevant that would help me for the goal of this first interaction with the client. I was looking for the ergonomic risk factors, but the engineering team obviously did not think about human factors. They were more focused on the technical specifications and product performance rather than how the user interacts with the product and how this affected health, safety, etc. This probably leads to their simplistic view of just getting the right seat design from a technical perspective.

When I performed a preliminary observation, I realised that my data intake form lacked many aspects of the assessment required. To my surprise, I was not allowed to take pictures on this visit due to commercial confidentiality reasons. Thus, I was dependent on sketchy drawings and line diagrams to draw up my initial format. During an ergonomic risk factors assessment, observing human–machine interactions is essential to understand and evaluate how tractor drivers operate all required controls without compromising safety and what could pose risks of developing musculoskeletal disorders.

I wish to note that when it comes to product evaluations, I am more of an intuitive user. I prefer to first try to use the product and understand its limitations and capabilities. However, on this occasion, I did not know how to drive a tractor, operate various controls, and use the added features required for farming. Nevertheless, just getting in and out of the tractor as well as sitting in a stationary tractor made me realise that there was more than just a driver's seat design to make this driver's seat comfortable without increasing low back pain discomfort. The most apparent factors I felt right away were ingress and egress on a high step and the awkwardness of reaching to the driver seat and important peripheral driver controls.

Before my visit ended, I had to prepare a preliminary presentation to the team about my risk assessment, its findings, and a recommended plan. I had to do all this before catching my return flight in two hours. As the scope was assumed to be only related to the driver seat, the team was expecting me to give a complete report on it. I was fortunate to demonstrate my findings while standing in front of the product, and I could give the team a visual demonstration to aid their understanding.

They seemed pleased with the demonstration with their heads nodding in agreement or understanding my explanations, but I was unsure if the client would move forward with the development of the new prototype based on the findings of an existing model. After all, I was talking about a long list of factors such as ingress/egress, control panel, design, visibility, etc., and not just seat design as anticipated by the engineering team. The client thought that all that might be needed would be a simple change of the driver's seat by giving it a certain shape or change of foam/cover, or similar. Thus, perceived problems and solutions by the team differed from my findings of various ergonomic risk factors that could influence driver's comfort. Simply put, my findings extended beyond the need for the redesign of the driver's seat.

I was left with the impression that a simplistic re-design of the seat was the solution expected by the team, and an exhaustive list of ergonomic factors that required broader consideration for re-design may not have been of interest to them. After I concluded my visit, I did not hear from the client for weeks, so I assumed that things would not progress. To my surprise, though, I got a call stating that we could go about the next steps in a phased manner. Later, I was informed that during their internal discussions while contemplating this project, they decided to hire an internal industrial product designer, rather than rely only on the engineering team.

Hence, in addition to the engineering team and me, the team included industrial product design specialists. However, we all were novices in designing a similar product prototype. There were no clear specifications or well-defined processes disclosed by the client or known to us. Therefore, it was a discovery journey for the team. Looking back at this initial project stage, having had some experience in similar product design would have helped to map the steps and milestones. I believe that we would have had a similar outcome in terms of the resulting product, simply because we followed a human-centred approach to the design, but we may have arrived at our endpoint faster and more efficiently.

Our first small win was the formulation of our problem statement:

- Users/drivers complained of the tractor seat being uncomfortable, causing low back pain.
- Need to improve the seat design of a new tractor prototype that was being considered as an upgrade for the new model.

The client's objective was to provide seat comfort for the tractor driver. How to go about it was the essential question for the product team. The client still believed that changing seats through minor modifications would improve comfort adequately. After our preliminary discussion, our primary goal was to assess the ergonomic risks from the perspective of the seat design as well as various peripheral driving controls and tasks (e.g. operating the clutch, breaks, accelerator, and steering wheels; and ingress and egress [entry and exit]). This would lead to a more complete assessment and minimise the risk of low back pain for drivers.

The challenge was that during my literature search on how to evaluate tractor seats and other peripherals, I could not find any standard formats or standardised tests that would suit our problem statement and design goals. Hence, I prepared and improvised on the required aspects and features necessary for the assessment and created my own ergonomic data collection format. The ergonomic analysis form was developed based on seat evaluation by the drivers, anthropometric measurements of seats, measurement of the location of various controls, and any obvious safety risk for the drivers. This format assisted me in the risk identification and communication with the team, ultimately aiding the decision-making.

This project made me realise that if standardised HFE tools do not exist or are not known, foundational design processes can be undertaken. For us, this meant that we reviewed our goals and objectives, and developed our own data collection instruments, while adhering to the tenets of human-centred design, such as involving subject matter experts and analysing human tasks. My simple but critical principle in the ergonomic assessment for product prototype and design is to incorporate everything that matters from the driver's perspective. A clinical occupational therapist needs to adapt when standard tools and formats do not match the purpose or objective of the evaluation. Standard tools at times do not help in a comprehensive view of ergonomic risk factors present or to fully manage the process of their prioritisation, directing resources, etc.

An inclusive data collection instrument that does not ignore any important system parameters is one of the most important aspects of ergonomic processes, as this is a starting point, the first step to make impactful changes. The ergonomic design process can be nuanced because of the context of equipment use and human tasks, so the design process must be agile, adaptable, and iterative (back and forth as learnings may happen). I feel that accepting the challenge with many unknowns and figuring it out while keeping in mind scientific principles/methods of human factors/ergonomics has worked well for me with one rule to follow: be very thorough in your evaluation.

Armed with my assessment, I helped the team understand my point of view on ergonomic factors, won their confidence, and eased my doubts of how to go about it. In the first step of the ergonomic assessment, the method used was an observational method coupled with job task analysis and anthropometric measurements. My assessment findings explained to the team were mostly in relation to tractor drivers' exposure to ergonomic risk factors leading to awkward posture, excessive pressure/effort, and contact stress. My initial observation list included aspects such as a high step and awkward reach to pull oneself enough to reach the driver seat; obstacles in the way to reach the driver seat; hard backrest with slippery seat cover; and high effort to push/pull the seat adjustment lever below the seat. The findings regarded the design and use of almost all system elements (e.g. steering wheel, hand accelerator, clutch, brake and accelerator pedals, gear and range shifter levers, parking brake, and control panel switches).

The referred objective of the ergonomic assessment was to come up with the right seat design for the tractor driver's comfort. However, that became a secondary objective for me. When we discussed the above, the next steps seemed a bit overwhelming. Where to start making these changes on the prototype considering the budget, staffing and time limitations? Thus, I first had to convince the team to prioritise safety concerns of ingress/egress of the tractor, climbing over the uneven floor of the tractor, and lifting legs to clear the floor when getting to the seat.

I was sceptical about the team's reactions to changing the course of action needed to make the product safer and comfortable. To my delight, the team was convinced and suggested looking at product design improvements in two phases. For the first phase, we decided to work on drivers' safety to ensure that they could get in and out of the tractor without hurdles and potential risks of trips, falls, and awkward postures leading to musculoskeletal disorders. Second, we contemplated seat design and drivers' interactions with the peripheral controls, which would be important for the comfort of the driver.

While everything seemed important, we had to pick the first line of defined and measurable changes that could be incorporated in the initial prototype of the design. My first presentation successfully demonstrated the safety risk of getting into the driver's seat which was related to high, uneven steps, poor-quality foot holdings, and gear lever on the floor obstructing driver's access to seat as well as sitting position, making driver's hip to remain in wide abduction while driving. These helped the product and engineering teams to understand the need to come up with engineering controls.

In the old model, the entry to the tractor was open using a high step, and there was no designated handrail to hold. Instead, the steering wheel and the side fenders of large wheels were used by operators to pull themselves into the tractor. Simply put, getting in and out of the tractor was no less effortful than climbing a big hurdle. Traditional farm tractors do not have doors! Hence, in this first phase of the new product design, we established and implemented some concrete physical safety parameters such as steps at the right height and with non-skid covering, adding handrails and relocating the floor gear lever. The engineering team's objective to engage with ergonomics prior to designing new prototypes gave them the most important aspect of a product: user safety first.

Based on the ergonomics standpoint, the engineering team created two steps with improvements in height from the ground, along with improved length and width required for full foot placement. A narrow high step extending outwardly out of the external profile of the tractor in the existing product was replaced with two steps. Table 8.1 presents some measurements on which the new prototype was based. The testing of these features by the team members suggested that the new design felt comfortable and secure. It would reduce the risk of trips, falls, and musculoskeletal disorders when adopting unnecessary awkward postures to get in and out of the tractor. Moreover, the fact that the length of the step was not extending out of the outer profile of the tractor reduced safety risks to pedestrians and other vehicles or when parking or driving on narrow and often barely motorable streets in villages.

Moreover, the flooring of the tractor was made of weatherproof black plastic material with a grooved texture. This seemed like an afterthought to protect the floor from all weather conditions because the floor was uneven. It seems like an ad-hoc addition to cover the floor. This uneven floor was a trip hazard. Most often, Indian truck drivers use open-toed footwear, which may get caught on the flooring. Given this input, the engineering team designed the floor space clear of obstructions, well-levelled, and weatherproof (i.e. India has tropical weather, and the tractors are

TABLE 8.1
Example of Design Changes of the Tractor Entry

Variable/Factor	Measurements		Comments
	Previous Model(mm)	New Product Prototype (mm)	
The first step from the ground	520	501	Less strain on the leg
Distance between steps	135	230	Improved stance
Inside width of steps	150	280	More surface area and stability for foot placement
Max. length of first step	253	290	More surface area and stability for foot placement
Length of step that is extended outside	152	None	Reduced safety hazard

heavily used in the rainy season when farming activities are at their peak). Also, in the new prototype, a grab bar of 100 mm length with a 10 mm addition of textured grip was installed. This would help to securely place hands and enter and exit out of the doorway to get to and out of the driver's seat without relying on wheel fenders and the helm.

Often, establishing priorities or changing the course of a product design prototype is a delicate matter because new designs constrain time and budget, and there can be limitations on skilled manpower resource allocation. However, despite being a novice product design team, we managed to assess the risks, set objectives, develop, and test prototypes. Witnessing the success of our first phase of the new product prototype was gratifying. This has now set us on our next phase of product development, and we were excited to further discover, problem-solve, test, and get approvals to progress further.

Regarding our next phase, the seat design, I had earlier realised that comfort was not just related to shape, size, contours, or quality of foam. It was much more than that. Through my experience and background in job activity and task analyses, I understood that the place you sit/stand on is as important as what one is doing while sitting. That human–machine interaction is part of "the essence of ergonomics". Even if the driver sits on the best-designed tractor seat, she/he will need to operate various peripheral controls and apply forces required to adjust seat levers and reach the hand accelerator, seat pedal, brake pedal, accelerator pedal, gear shifter lever, operate various types of switches, and adjust mirrors to ensure visibility.

Indeed, the specific seating characteristics (e.g. length and height adjustment, shape of backrest and seat, or quality of foam covering) still mattered for the driver's comfort due to the required forces to operate some controls while sitting. The location of the gear control lever in the middle of the legroom affected the normal sitting position of the driver. The driver had to sit with wide open legs almost towards the end range of hip abduction. Hence, to make the driver seat comfortable, we needed to address most of these issues. I faced challenges in articulating what would be needed and explaining an exhaustive list of features required for comfortable sitting while operating the tractor and performing various farming operations.

Nonetheless, following the initial success of demonstrating the required product design and having a successful working prototype, the team was open to receiving and understanding the requirements. We started with a focus on the first level of driver seat improvements. By using a "look, feel, and fit" approach, we consulted with the anthropometric data we had collected and made improvements related to seat height, seat pan depth, backrest design, and armrest height. Apart from the static components of the seat, we also worked on dynamics controls for seat adjustments, spring tension, and the length of levers, as well as the addition of a driver seat safety belt that was missing in the existing tractor model.

The team was pleased with the improved features but understood that this was not sufficient for the established goal. The goal was to achieve adjustable seat features for a better sitting experience while operating various controls. Additionally, the next step was to work on minimising awkward reaches and difficult grips of peripheral controls. For example, the gear lever with top knob moulded to keep grip in the wrist neutral position was moved to the right side of the driver seat for ease of operation (i.e. instead of the current location being in the middle of the driver's legroom area). This minimised awkward postures with excessive reach and forward bend as per the operations in the previous model. Next, we would discuss as a team and agree upon the set of changes to be added for the next prototype assessment. It took several months to get the display unit, clutch, accelerator, brake pedal, hand control, steering wheel, position control, etc., ready for review by the entire team. The new, improved version of the prototype was developed, tested, and retested against the established ergonomic design objectives.

Despite all the doubts when I started on this project, the engagement with a novice team and the journey of working through various milestones led us to generate a prototype that resulted in a new prototype of a tractor. In my role as an ergonomic expert, I learned that looking at all details of a product and appreciating the multiple facets of interaction between humans, equipment, and tasks involved leads to a thorough and inclusive evaluation. This, in turn, offers a complete understanding of ergonomic risk factors and supports the process towards the generation of solutions, which, when accepted by the team and stakeholders, lead to successful outcomes.

Armed with the success of this prototype and the launch of the new model in the market, the team aspired to continue with more improvements in new products launched for different categories of farm activities. There was a market demand for better, improved models to support the economically changing conditions that included improved purchasing power of Indian farm owners. Thus, the company leaders wanted to move forward with the next improved and state-of-the-art tractor. This would have a proper enclosed driver cabin, be more comfortable with climate control options and highly adjustable seat options for maximising comfort. They wanted a tractor that would contribute to the least possible fatigue. Following this vision for a more advanced tractor model, we had multiple meetings.

I tried to help the team understand ergonomics in greater depth, especially for features such as designing closed driver cabins with climate control options. Also, to improve the next tractor model, we needed consumer feedback to sense whether the competitive Indian farmer would be receptive to a more advanced model. This would require extensive research and considerable effort. Once those prerequisites were understood by the team, they realised that to implement their vision of state-of-the-art tractor, greater effort was required along with bigger investments, advanced skillsets, and increased staffing.

Admittedly, the next product was a more complex problem to solve. Also, the company realised that in-house ergonomic experts would be required to build a new prototype. Although I did not get to work on the next model, still there was a big win. The company started considering ergonomics as an inextricable part of product design! That was a huge success as it would lead to a better quality of future products and benefits for the end customers, the Indian tractor drivers. As a farmer's daughter whose revenues depended on farming and the effective use of tractors, having the opportunity to help drivers through ergonomic tractor design will forever remain a great experience for me. Moreover, becoming part of a team with multiple disciplines pushed me to improve my knowledge in other areas and work with everyone to realise the new product design.

When looking back, I believe that I should have first worked on developing a complete vision of the design and the steps/phases involved. I think that this would have made it easier for the team because they initially thought that it would be only a one-step assessment to generate redesign recommendations. However, ergonomic design is always challenging. Whatever design you are working on, the context of the equipment, the users, and their tasks and needs must be considered using scientific methods.

Moreover, developing measurable and concrete milestones broken down in several phases of product development would be improvements that I want to implement in future projects. Also, I would like to know beforehand the roles and responsibilities of all stakeholders involved in the design process. Quite often, there were new members and new roles added which made it difficult at times and impeded the speed of the process because, each time, we had to review the project and help the new members understand the design process and objectives.

Successful outcomes can be achieved, and harmonising team efforts is "doable" when we connect the dots of our diverse experiences, and we remain open to learning from our past work projects.

9 Off-The-Road Tyre Management

The Good, the Bad, and the Ugly

Paulo Gomes
Segurança Diferente

CONTENTS

If you had to think about a high-risk activity in mining, what would come to your mind? Mining blast? Perhaps operators of those large trucks? Think again. According to Rasche (2019, 6:15), off-the-road (OTR) tyre fitting is so risky that people working with tyres, tyre fitters, "[...] are between 10 and 12 times more likely to be fatally injured than a mine workshop maintainer fitter". Amongst mining maintainers, tyre fitters are often seen as low maintenance workers within a workshop hierarchy. The *tyre bay*, where tyre maintenance occurs, is often an improvised work area normally located in the corner of a workshop.

If someone wants to become a tyre fitter in Australia, they will be required to complete a formal training called *AURKTJ011 Remove, Inspect and Fit Earthmoving and Off-The-Road Tyres*.[1] In theory, this training should provide the trainee with sufficient information so that they can identify and manage all critical risks related to removing, inspecting, and fitting earthmoving and OTR tyres. However, to my knowledge and experience, the training is mostly conducted in class, supported by slides, and with little opportunity to practice. When some practice happens, it is typically with small-scale models of a wheel and tyre to match the environment of a training workshop. These limitations could be attributed to a large number of types of OTR wheels and sizes available, making it extremely difficult for registered training organisations to exactly represent the work conditions that tyre fitters will face on a mining site.

---

[1] https://training.gov.au/Training/Details/AURKTJ011

DOI: 10.1201/9781003349976-9

Assembling a wheel and fitting a tyre are complex tasks. Several original equipment manufacturers (OEMs) and different designs of OTR wheels are in use in the mining industry. Some large sites operate with almost all types of OTR wheels, including single-, two-, and multiple-piece wheels. Large-haul trucks, those massive vehicles that normally carry a great volume of ore from the mining pit to the dumping area are usually fitted with OTR wheels composed of multiple pieces, and they are fitted with the largest tyres in the world. These tyres operate under such extreme internal pressure that they are described as 'moving bombs'. Something can easily go wrong so that the violent pressure release can transform any part, including a small piece of rubber, into a projectile. Surviving an accident with those tyres is nearly impossible (Taylor, 2010).

Over the years, research investigating haul-truck improvements has been focussed mainly on the performance of the engine and carrier capacity of the truck. Different OEMs have developed their own OTR wheels, which contain several components that seem to be similar, but because of slightly different angles, cannot be mixed If components from different OEMs are mistakenly put together, or assembled in the wrong order, a catastrophic disassembly of the wheel assembly can happen. More specifically, the lock rim, rim base, and bead seat band are the three main components of a wheel that have killed people when assembled wrongly. Despite several fatal accidents involving mismatching wheel components, the design of OTR wheels has seldom changed for almost half a century. What is worse, the *AURKTJ011* training does not mention anything about these parts or the risks if they are assembled wrongly (Rasche, 2019, 20:31).

There are a high number of possible combinations when assembling a wheel, compounded by time pressures which can contribute to hazardous working conditions. Having a truck stopped for heavy maintenance is one thing. The other is having an unproductive truck because of a problem with the wheel or tyre. In addition, several tyre changes are performed at night, in the dark, with poor visibility of the tyre and parts. Components are also required to be properly cleaned to remove any signs of rust, aggravated by the possibility of mixing components from different OEMs. This is a receipt for a disaster.

In recent years, significant improvements have been noticed in Queensland (QLD) and Western Australia with the release and enforcement of tyre management standards and guidelines by their respective Mine Departments. Nonetheless, critical hazards to the activity such as thousands of possible assembling combinations given the number of components of a wheel, the lack of standardised wheel design, and poor identification of the parts are well documented and known. There is an impracticable expectation that each tyre fitter will manage these risks on the job (Hassall & Boyle, 2016). In addition, the lack of conveying critical information in some training programmes and the working conditions of a tyre fitter remains unaddressed.

Against this background, I share two cases that convene my experience when I worked for a large Australian OTR tyre management company overseeing their operations across QLD, New South Wales, and New Zealand. Given my background in working as a safety specialist in mining sites around the world for several years, when I joined the company I thought that I knew enough about mining and tyre management. However, as the stories will tell, I could not be more wrong. Not long after starting the new job, I realised that the situation was more complicated than anticipated.

The first case is about success, creativity and appreciative inquiry (Whitney et al., 2010). One of the first things I proposed when I started the new job was the introduction of a new way of collecting safety information from the tyre fitters. Rather than hazard reports or behavioural observations I proposed to talk directly to the fitters. Further to collecting information about what can go wrong, I decided to open the space for people to share insights to improve the safety and productivity of their work (Provan et al., 2020). Therefore, in the following section, I describe how the work of the tyre fitter improved with the ideation, design, creation and implementation of a new tyre fitting tool.

However, life is not only made of good stories. Experiences that do not go as expected can also teach us a lot, or even more, as some say. In the second case, I am presenting the dark side of technical standardisation. The Regulator, other stakeholders, and I tried to push for a change to the Australian Standard 4457 on OTR Wheels and Tyres (Standards Australia, 2007). Little did I know about the politics involved in amending standards. The mining industry is powerful and can lobby against safety initiatives whenever the changes to regulations or standards increase the costs or impose additional hurdles to their businesses. At the end of this story, I reflect on the lessons learned on the 'battlefield'.

LET'S EMPOWER THE END-USERS TO FINISH THE DESIGN

This first case relates to when I decided to change the culture of my organisation around 2015. Back then, the focus was almost entirely on managing numbers imposed by clients, mainly large mining operators, who were interested in measuring the number of hazard reports against actions completed. However, 'the number of hazard reports done tells us nothing about their quality' (Smith, 2018). There was no space for consultation or contextualisation because they believed that workers only needed to follow their procedures to be safe. That was not surprising as several mining managers believe that most accidents happen because workers either did not follow a procedure or there was no procedure to follow (Laurence, 2005).

Incredulous with the context, I decided to introduce the appreciative inquiry approach to engage the workforce to identify successes and strengths that, in turn, would move the organisation to our desired future. The term 'appreciative inquiry' was first coined by David Cooperrider in 1980[2] and aimed to serve as a problem-solving tool based on the positive aspects of the organisation rather than the negatives. Four steps drive the appreciative inquiry method, commonly referred to as 4Ds (Cooperrider et al., 2008):

- **D**iscovery, where the best of previous and current experiences is discussed with participants, exploring what motivated the inquiry.
- **D**ream, when participants are asked to imagine their group, department or organisation at its best.

---

[2] https://www.davidcooperrider.com/ai-process/

- **D**esign, where participants are invited to describe what the organisation should be with provocative propositions.
- **D**estiny, the final step, aims to evaluate the proposed ideas and implement them.

Applied to safety, appreciative inquiry is conducted as a focus group, in the workplace, with different frontline workers from a particular operation participating in the session. They are invited to discuss the process, procedures, tasks, activities, and work conditions in general (Discovery). Normally, the initial part of the session leads to the acknowledgment of risks, actual practices and challenges in the ways of working. Next, they are asked what the ideal work and its setting would look like (Dream). Then, they are invited to share ideas of new ways to do their work (Design). Using the information shared, the safety specialist and the engineering department can translate the needs, dreams, and proposals into practical solutions to be implemented in operations (Destiny).

In one of the appreciative inquiry sessions I conducted, a Senior Tyre Fitter shared that changing tyres on an EH5000 haul truck was riskier than most of the other haul trucks. The Senior Tyre Fitter reflected on one day when a new EH5000 haul truck was delivered to the mine site he was working. Five fitters from the truck OEM were required to assemble the wheels containing a dozen large steel pieces called spacers. The spacers must be positioned inside pockets around the wheel drive to secure the wheelbase. Given the intricate aspects of fitting a dozen spacers, the OEM fitters used a large and bulky tool suspended by a Franna crane. Two fitters supported the tool above their heads while trying to align each spacer to fit into the pockets. This activity continued for half a day. The Senior Tyre Fitter observed the task and listed several things that could have gone wrong with the work observed, some of which could lead to fatal injuries (e.g. hands being caught between the tools and the wheel and uncontrolled movement of the suspended load).

The Senior Tyre Fitter recounted that he spent days thinking about how this task could be done differently, safer, and more efficiently. He then selected a couple of steel tubes, adapted a crowbar and created a proof of concept of the ideal tool to help with fitting the spacers. However, because he was not sure whether the tool would be well received by his colleagues and supervisors, and because sharing new ideas in the past was deemed risky and improvisations were reprimanded, he kept everything to himself, under the workbench in the tyre pad.

Fortunately, during the appreciative inquiry session, the Senior Tyre Fitter revealed the tool, explained the rationale, and provided detailed instructions on how to use it. I was hooked. Still, on the spot, the session participants brainstormed the concept, worked on a couple of technical drawings of the tool, and defined some requirements, including that it had to be light, strong, with no sharp edges, and easy to manoeuvre. I then promised to take the idea to the office and evaluate the feasibility of creating a prototype of this tool.

The safety department did not have a budget to pay for the prototype. Moreover, the safety manager would not approve it because one of the organisation's most profitable projects had just come to an end. I then decided to contact an engineering company that manufactured a couple of work platforms for us. I asked if they could, as a favour, use scrapped steel pieces to put together a rough prototype of the tool so that it could be used on trial. The company agreed!

Once the rough prototype was ready, a first qualitative risk assessment was completed for the use of the tool, and we concluded that it did not introduce any side risks. Then, the idea was presented to the organisation's management team and the client (i.e. the mining operator) on site. The client was fascinated by the idea and requested to observe the trial. Even with the rough prototype, the task did not require a Franna crane, eliminating the most critical risk from the task. It could be safely performed by one person (i.e. not a team!) holding the tool below shoulder height, and the task was completed in approximately three hours (i.e. half of the time!). The client was so impressed that they offered to pay as much as it would cost to develop a proper tool. Three prototypes were developed and tested in a period of six months until everybody involved in the project was happy with the final design. The final tool is comprised of a set of three clamps made up of stainless steel, and it comes with a user manual. The tool was so relevant to the tyre fitter work that it was shared with ten other mine sites owned by the organisation in Australia and overseas.

This story teaches us that while safe work may be an objective, task redesign usually leads to improved efficiencies. It shows that listening to the workers and the frontline people should be one of the main goals of leaders who want to positively impact the workplace. Our workers know the problems, and they could also have the best ideas for the most suitable solutions. As important as it is to listen to and engage with workers, changing the work conditions should also be our priority since we have access to more resources than a shop floor worker. The solutions should be profound and should start small, with a concept or prototype to be validated in the first place. Then, a small-scale pilot validation should be conducted to evaluate the feasibility and any additional risk added by the solution. Once the prototype is validated and new risks are revealed and properly controlled, the tool can be implemented widely. However, it is important to keep in mind that the solutions are never perfect and require continuous adjustments over time based on feedback from real-world implementation.

In 2017, this tool redesign project was submitted to the National Safety Awards of Excellence in Australia and won an award in the category of Best Individual Workplace Health and Safety Achievement (Godwin, 2020).

CHANGING A STANDARD REQUIRES MORE THAN GOOD WILL

The fitter tool explored in the previous section, despite being incredibly effective, still does not prevent the potential risk of mismatching components. Several OTR tyre-related fatal accident reports and most of the tyre fitters with whom I worked agreed that trying to assemble a wheel with unidentified components or components from different manufacturers is the most critical safety concern in the industry. Indeed, tyre handling and other related tasks also hide risks (e.g. when using equipment). However, of concern in this instance is that a small mistake during the assembly process can result in a catastrophic disassembly of the wheel under approximately 150 psi.

In general, mismatching components is a significant concern because many people stand in front of the tyre several times a day. Any mistake can kill them or somebody nearby. The tyre fitter's worse nightmare is to work at a mine site operation with several different OTR wheels from various OEMs and be tasked to change the inner tyre on a dual wheel assembly while working alone at night, which is very common across mine sites around the world. Not only are tyre fitters exposed to the consequence of this hazard but could also be blamed for mismatching components.

As Brady (2019, p. 29) pointed out, *'a common view in the mining industry is that "human error" plays a substantial role in fatalities [... and] a common term used in the industry is "lapses in concentration"'*. This can lead to the unfair assumption that tyre fitters choose to get hurt and consciously lose concentration, or they lapse in concentration, whereas, in reality, the bad design of the wheel and poor working conditions set the tyre fitters to fail. All these are further aggravated by the fact that their training focusses on the minimum necessary level of competency that, however, might not match the level of complexity required by the job demands (decision-making in limited time; basic risk assessment, analysis, and treatment; planning, etc.).

Following a meeting with some engineers and the safety team on this matter, I left the room frustrated because there seemed to be almost nothing that could be done to prevent mismatching components. Even the organisation's idea of marking and identifying OTR wheel components could not go ahead because of the potential damage to the steel, there was a high chance that any attempt to mark critical wheel components could result in brittle points and, thus, cause mechanical failure. The seemingly good solution to mark components could become a worse problem.

Still determined to do something to improve safety by design, I found a QLD Mines Health and Safety Inspector on LinkedIn who had published several papers on critical risks related to OTR wheels and tyres. After a quick exchange of messages, we went for a coffee and discussed some ideas on what could be done to assist in improving safety in the OTR tyre management industry. We realised how much we shared in terms of frustrations and the desire to do something more tangible to improve the work conditions of tyre fitters. QLD Mines Department had also done as much as they could to improve the OTR tyre management industry from the regulatory perspective, particularly by issuing the *Recognised Standard 13*[3] and the *Guidance Note 31*[4] *for Tyre Wheel and Rim Management*.

Both documents had a significant positive impact on the mining industry in Australia as they provided mining operators with guidelines on how to comply with minimum safety requirements. However, the design of wheels was still an inherent risk that could only be addressed if the stakeholders targeted a more influential reference, the Australian Standards (AS). The *AS 4457.1 Wheel Assemblies and Rim Assemblies*[5] was last updated in 2007. It was supposed to be supported by a committee of experts to collect good practices and feedback from the industry and keep the document updated. However, the committee was dismantled circa

---

[3] https://www.resources.qld.gov.au/__data/assets/pdf_file/0004/986071/recognised-standard-13.pdf

[4] https://www.resources.qld.gov.au/__data/assets/pdf_file/0005/1407785/qld-guidance-note-31.pdf

[5] https://www.standards.org.au/standards-catalogue/sa-snz/mining/me-063/as--4457-dot-1-2007

2012, and since then, although two coroner's tyre-related fatal accident reports in QLD urged for the *AS 4457* to be updated, nothing has been resolved (Barnes, 2006; O'Connell, 2014).

Soon after our meeting, the inspector and I formed a working group with tyre management experts from all major OTR wheel OEMs and two representatives from mining powerhouses. Our intention was to standardise the design of the wheel components, in particular the angles. We also planned to develop a guideline for marking and numbering wheel components to assist tyre fitters in visually identifying and selecting matching wheel components. The idea was well received by the group. Everyone was feeling confident that they could finally solve a problem that had remained unresolved for a long time. We worked together for months until the proposal to update the Australian Standard 4457.1-2007 was completed.

The proposal was then submitted and rejected for the first time at the end of 2017 and then again for the second time. The team that assessed the proposal told our group that there was no consensus among the members on the necessity for the *AS 4457.1* to be reviewed; no further details and explanations were provided. Despite the consensus among the experts of the OTR tyre management working group that changing the Australian Standard was the only solution for the problem of mismatching wheel assembly components, the team appointed by Standards Australia did not provide us with any strong arguments to justify their decision to reject our proposal. There was something very wrong here.

Unfortunately, given that no details were provided about the rejection of our request to modify a 15-year-old standard, we could only speculate on what happened. Significant changes in the mining industry usually occur only if they are to address something that can cause serious damage to the reputation of organisations or is imposed by investors. For example, after two major tailing dam failures in Brazil, a group of international investors was concerned about the legacy issues related to the disposal and closure of tailing dams. This forced the International Council for Mining and Minerals to lead a working group responsible for the development of the Global Industry Standard on Tailings Management in August 2020 (Hopkins & Kemp, 2021). Otherwise, many more Brumadinho-like dam disasters (Rotta et al., 2020) could have happened across the world.

Admittedly, at the time of writing my chapter, March 2022, the *AS 4457.1* standard is shown as pending revision. However, I am not aware of what changes will be incorporated. Nevertheless, there is no doubt that the solution for mitigating the risk of mismatching components requires a substantial investment from the industry because several OTR wheels would have to be phased out and replaced by better and safer wheel assemblies. This is a significant undertaking, indeed. However, as an accident due to a mismatch can affect only one operator and is unlikely to make cover page headlines, this might not hold political sway to effect change.

As safety professionals, we might underestimate the capacity of industry to ignore serious problems by using their well-known blame-and-shame way of business to push things under the carpet (Pitzer, 1999; Gunningham & Sinclair, 2009; Tian et al., 2014; Stemn et al., 2019). Ineffective safety actions, such as hanging 'safety' signs on the walls and carrying out a couple of toolbox talks a month, are valued practices that do not impact operations in an effective manner (Dekker,

2017). I also underestimate the political and power-related interactions that happen in any industry. I am generally good at working in the sharp end but bad at playing the blunt-end game.

IN HINDSIGHT

The first case demonstrated the importance of empowering people to expose their ideas in a supportive and psychologically safe environment. It is easy for us to jump to conclusions straight away and move on to the next issue to be addressed. However, sustainable solutions can only be created if we pay attention to what workers need to do their work safely and more effectively.

The second story presented a real-world example that good intention to solve critical risks is not enough. Changing an outdated standard was revealed to be more challenging than anticipated (and nearly impossible). This is true when there are more interests on the line other than making the industry trustworthy and safer. With the benefits of hindsight, proposing a working group inside Standards Australia may have been a better strategy. That seems better than the alternative to do nothing and sit and wait for a major accident to happen so that the standard can be updated in blood. Only time will tell.

REFERENCES

Barnes, M. (2006). *Inquest into the Death of Peter Whitoria Marshall.* Office of State Coroner.

Brady, S. (2019). *Review of All Fatal Accidents in Queensland Mines and Quarries from 2000 to 2019.* Department of Natural Resources, Mines and Energy.

Cooperrider, D. L., Whitney, D., & Stavros, J. M. (2008). *Appreciative Inquiry Handbook* (2nd ed.). Crown Custom Publishing.

Dekker, S. (2017). *The Field Guide to Understanding 'Human Error'* (3rd ed.). CRC Press. DOI: 10.1201/9781317031833

Godwin, H. (2020). *Mining Supervisor Champions Safety.* Otraco. https://otraco.com/news/mining-supervisor-champions-safety

Gunningham, N., & Sinclair, D. (2009). Regulation and the role of trust: Reflections from the mining industry. *Journal of Law and Society,* 36, 167–194. DOI: 10.1111/j.1467–6478 .2009.00462.x

Hassall, M. & Boyle, M. (2016, August 14–17). Addressing Tyre Risks with Critical Control Management – A Collaborative Industry Project [paper]. A past forgotten is a future repeated. Health and Safety Conference 2016 of the Queensland Mining Industry, Gold Coast, Australia.

Hopkins, A., & Kemp, D. (2021). *Credibility Crisis: Brumadinho and the Politics of Mining Industry Report* (1st ed.). McPherson's Printing Group.

Laurence, D. (2005). Safety rules and regulations on mine sites – The problem and a solution. *Journal of Safety Research,* 36. DOI: 10.1016/j.jsr.2004.11.004

O'Connell, D. (2014). *Inquest into de death of Wayne Macdonald.* Office of State Coroner.

Pitzer, C. (1999). New thinking on disasters: The link between safety culture and risk-taking. *The Australian Journal of Emergency Management,* 14(3), 41–50. DOI: 10.3316/ielapa.392186186255759

Provan, D. J., Rae, A., & Dekker, S.W.A. (2019). An ethnography of the safety professional's dilemma: Safety work or the safety of work? *Safety Science,* 117, 276–289. DOI: 10.1016/j.ssci.2019.04.024

Rasche, T. (2019, August 19–22). A Study to Investigate Uncontrolled Tyre and Rim Disassembly [video presentation]. Working to the future, Health and Safety Conference 2019 of the Queensland Mining Industry, Gold Coast, Australia. https://qmihsc2019. evertechnology.com/conference-session/cross-border-safety-opportunities/

Rotta, L. H. S., Alcântara, E., Park, E., Negri, R. G., Lin, Y. N., Bernardo, N., Mendes, T. S. G., & Filho, C. R. S. (2020). The 2019 Brumadinho tailings dam collapse: Possible cause and impacts of the worst human and environmental disaster in Brazil. *International Journal of Applied Earth Observation and Geoinformation, 90*. DOI: 10.1016/j.jag.2020.102119

Smith, G. (2018). *Paper Safe: The triumph of bureaucracy in safety management*. Wayland Legal Pty Ltd, Perth.

Standards Australia. (2007). AS 4457.1-2007: Earth-moving machinery – Off-the-road wheels, rims and tyres – Maintenance and repair, Part 1: Wheel assemblies and rim assemblies.

Stemn, E., Hassall, M. E., Cliff, D., & Bofinger, C. (2019). Incident investigators' perspectives of incident investigations conducted in the Ghanaian mining industry. *Safety Science, 112*, 173–188. DOI: 10.1016/j.ssci.2018.10.026

Taylor, G. (2010). *Integrity Testing of Earthmover Rims. Mines Safety Bulletin No. 103 (Version 1)*. Queensland Mines Inspectorate, Resources Safety & Health Queensland.

Tian, S. C., Kou, M., Sun, Q. L., & Li, L. (2014). Analysis of near Miss reporting will in China coal mines: An empirical study. *Advanced Materials Research, 962–965*, 1127–1131. DOI: 10.4028/www.scientific.net/amr.962-965.1127

Whitney, D., Trosten-Bloom, A., & Cooperrider, D. (2010). *The Power of Appreciative Inquiry: A Practical Guide to Positive Change* (2nd ed.). Berrett-Koehler.

10 The Human Factors Practitioner in Engineering Contractor-Managed Investment Projects

Ruud Pikaar

ErgoS Human Factors Engineering

CONTENTS

DOI: 10.1201/9781003349976-10

Human Factors/Ergonomics (abbreviated to HF) is always design-driven. Dul et al. (2012) summarizes this by defining HF as (1) the scientific discipline concerned with the understanding of the interactions among humans and other elements of a system and (2) the profession that applies theoretical principles, data, and methods to design to optimize human well-being and overall performance. HF practitioners apply their knowledge in technical and organizational settings. They interpret and integrate the results of scientific research into the design of complex human–machine systems.

At our HF Engineering company, we have a large, although unpublished, database of nearly 40 years of HF projects related to the process industry, oil and gas, manufacturing, rail and road traffic management, and offshore shipping. In this chapter, we present a case of a new Floating Natural Gas processing system. It is a huge vessel, which cannot navigate on its own. It holds several process units, liquified gas storage, and loading facilities. The process units are supervised and controlled from a control centre, which looks much the same as onshore control centres in the oil and gas industry.

This project was one of eight oil and gas control centre projects carried out within comparable organizational settings (Pikaar et al., 2021). Three of these projects regarded liquified gas processing and were situated with the same Engineering Contractor. The project of focus in this chapter was the second one and will be referred to as the "Project". Experiences from the first project will be mentioned, including our learnings for the next project. I use the term "we" frequently, as this type of project is always carried out by a team of several HF consultants of our company. The team was the same for all three projects.

The project owners (referred to as "Company") were major international oil and gas companies. The projects were managed by an Engineering and Procurement Contractor (EPC). For the first two projects, our company, which employs five registered senior European Ergonomists (Eur. Erg.), was subcontracted by the Instrumentation and Control System sub-Contractor (ICSC). For the third project, we were contracted by the EPC. The Project was purchased at a fixed price, based on approximately 100 days of HF work.

STORIES TO BE TOLD

There are many stories to be told about the Project. I selected several typical stories about the balance between "good HF practices" and commercial engineering issues. What is acceptable for the HF professional? What to do if it is not good enough or even a bad design, seen from an HF point of view?

Our stories are organized along five project phases:

1. *Scope Definition:* The scope of work is developed before you commence an HF consultancy contract.
2. *Kick-off and Task Analysis:* The kick-off concerns the organization of the HF contribution, as well as an explanation of what HF is about. The next step should be a task analysis.

3. *Disappointment Phase – Design Review:* Based on task-related knowledge, the initial design will be reviewed. A review inevitably leads to HF shortcomings.
4. *Bittersweet Phase:* HF-based improvement design proposals, presented in 3D and through Virtual Reality.
5. *Document Control:* Wrapping-up documents and close-out.

PROJECT PHASE 1: SCOPE DEFINITION

Communication between the EPC or the ICSC and the HF consultant starts with a Request for Quotation (RfQ) or Invitation to Bid. Usually, RfQ is not written by HF experts. RfQ is approved by the Company and accepted by the EPC and ICSC. At the table of the ICSC-buyer, the RfQ-specified scope of work is hardly negotiable because changing a Company-approved document is very difficult. If you want the contract, you need to comply or at least say that you will do so.

The main interest of the customer is in workplace design and environment. However, to arrive at an acceptable workplace design, the HF consultants need to consider operator tasks and workload. One needs to determine or validate the number of operators and consoles required for safe operations. This will also tell you how many process graphics are to be expected, as well as the number of screens that would be needed in the workplace (Pikaar, 2012).

Review of Initial Design

For our cases, the RfQ requested a review of the initial control centre design, including solving any ergonomic problem arising from this review. The initial design is the result of the so-called Front-End Engineering Design (FEED) phase at the EPC. In our cases, there had been no HF input during FEED. EPCs tended to rely on copy–paste from earlier projects and, if available, Company ergonomic guidelines. If we were to find several ergonomic design-related issues, there would be lots of work to do. Remember these projects are contracted at a fixed price. Getting an additional budget is difficult and time-consuming. Hence, it may be tempting not to report all issues arising from a review.

For the first project, although promised otherwise, it turned out that there was no layout drawing of the control centre. The control room (CR) area was an empty space. Instead of doing a review, we ended up with a full design cycle. This is not necessarily bad because you can introduce a splendid ergonomic design. However, it may take additional time. Learning from this, we offered more days of Project work for the review. Fortunately, in this case, a conceptual design of the room layout was available, however, not based on any ergonomic guidelines.

What Is an Ergonomic Study?

Usually, the RfQ requires the deliverable "ergonomic study". In the first project, there was no further description of "ergonomic study". During clarification, we suggested referring to the HF design steps specified in part 1 of ISO 11064 *Ergonomic design of control centres* (1999–2013). Surprisingly, for the Project, the RfQ included this suggestion. We considered this to be a small success!

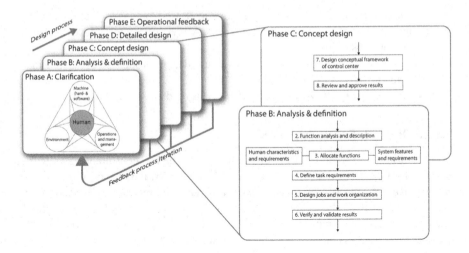

FIGURE 10.1   Five phases of HF design. (Adapted from ISO 11064-1.)

Part 1 of ISO 11064 specifies in section 5 a "Framework for an ergonomic design process of control centres", consisting of the following phases (Figure 10.1):

A. *Clarification* of purpose, context, resources, and constraints of the project.
B. *Analysis and Definition:* This includes a functional analysis and allocation, job design, and validation of obtained results. Unfortunately, the text doesn't clarify how and when a task analysis based on interviews and observations of operators should take place. In ISO 11064-Part 2, such a task analysis is presented as being mandatory because a good insight into operator tasks is essential for any design.
C. *Conceptual Design:* This consists of two steps: (1) the design of a conceptual framework for the CR and (2) review and approval of the conceptual design. The design concerns CR layout, routing, workplace layout, instrumentation, a communication matrix, visual tasks, etc.
D. *Detailed Design* of operator consoles, fixed and loose furniture, arrangement/mounting of instrumentation, cabling, and process graphics. This step also includes the specification of environmental requirements.
E. *Operational Feedback* for evaluation purposes after commissioning.

Notably, it is a common practice that process graphics/interaction design is not in the HF scope. ICSCs are convinced of being competent in this HF aspect, but often they do not comply with basic interaction guidelines (Pikaar, 2012).

Also, during the development of ISO 11064, I recall significant discussions and some confusion about Phase B *Analysis and definition*. The interpretation of "task analysis" is ambiguous. Does it include the allocation of tasks or is it merely acquiring knowledge about actual tasks? Functions and function allocations are defined on a higher level of abstraction and may be related to determining the level of automation of the production system.

Anyhow, in practice, the use of functions instead of tasks doesn't work for an engineering team having a limited understanding of HF because it is too abstract. Therefore, we always start with a task or situation analysis, meaning what is happening in the existing or a comparable control centre. Next, we draw conclusions and specify requirements for operator jobs and work organization in the new system. These requirements are input for Phase C. At this point, we perform a task allocation to design jobs and, thus, workplace requirements to accommodate those jobs. Still, this is not an easy approach because technical engineers and project managers usually expect that an HF study is, would, or should be limited to Phase D *"Detailed design"*.

What Did We Learn from Scope of Work Issues?

The introduction of the ISO 11064 five phases was successful. It landed in the RfQ for this and other projects. Did this approach work for the Project? Hardly, because it remained difficult to convince the engineering team of the necessity of a systematic task allocation, work organization design, and job design because *"Operator jobs are the Company responsibility. Just give us a detailed workplace design specification"*.

In the first project, two experienced former shift supervisors participated in the project team. In the Project, two commissioning coordinators participated. In both cases, they were our source for task-related information and provided valuable feedback on our proposals and recommendations. They reported to the Company, not to the EPC. Consequently, we could not support them during meetings at the Company level in convincing management. That was a task in which they didn't always succeed. Commitment to HF is a key issue, particularly at project lead and higher management levels at an EPC and the Company. Nowadays, we put the commitment issue high on the agenda of the RfQ clarification and the kick-off meetings.

PROJECT PHASE 2: KICK-OFF AND TASK ANALYSIS

The project team consisted of an EPC project lead, the person responsible for the ICSC project, two HF consultants, individuals from several engineering disciplines, and several Company representatives.

Kick-Off

The kick-off meeting was organized in three half-day parts. Part one was dedicated to getting to know the project team. It included an explanation of HF guidelines, design approach, and scope of work. Part two was dedicated to acquiring knowledge of the processes to be supervised and controlled, operator jobs, and work organization. For the Project, we had to rely on our own understanding of the process units because for this greenfield project there were no experienced operators available. Justifiably, the Company might have considered it too early to think about staffing as it would take another three years before the vessel becomes operational.

The third part of the meeting regarded a lengthy explanation of the document control principles and how to use the document issuing software. During the presentation, we did not really understand the complexity of the document control system. During the project, we relied on the ICSC for assistance. Notably, during all progress meetings, presence varied. Many participants were only available when their own specialism was tabled. Consequently, it was difficult to share our HF expertise effectively and reach a consensus on the integration of design inputs from different disciplines.

Operator Workload and Work Organization

We were concerned about the allocation of process units to operator positions. However, the project team refused to understand the importance of task analysis for estimating operator workload. We managed to get an overview of the operator roles and work organization. Operators would be trained for only one of the following work positions:

- 1 operator + 1 supervisor for subsea production.
- 2 process operators + 1 supervisor for compressor units and utilities
- 1 marine operator + 1 supervisor for offloading and all marine activities

We made workload estimates based on our expertise. As a result, we expected overload problems for the marine operator, amongst other things, because this person had to handle two very different types of instrumentation: maritime instruments and process supervision. Also, the work organization looked slightly odd. One supervisor for each operator position! So, what would be the supervisory role? Our concerns were discussed, and some changes were effectuated in the allocation of process units to work positions.

What Did We Learn about the Kick-Off Meeting and Task Analysis?

It should be clear who is responsible for leading the team, running meetings, taking minutes, and drawing conclusions and/or making decisions. At first, the EPC lead had no intention of chairing the meeting and left it to us. That did not work. Being a sub-sub-contractor, you don't have much impact on the format and outcomes of a meeting. Also, being chair and designer/consultant at the same time is difficult. In the Project, the ICSC responsible for the HF contribution took over, which was helpful and had a better impact. The ICSC has better possibilities to escalate issues. A minor drawback was caused by several participants speaking a native language other than the project language. Finally, the Project strengthened our idea that one needs a certain amount of understanding of the technical processes involved to fill in the blanks about operator jobs and determine role demands.

PROJECT PHASE 3: DISAPPOINTMENT PHASE – DESIGN REVIEW

The CR was a rectangular room (9×21 m²), with four operator consoles. Corridors were projected on the left and right sides as indicated in Figure 10.2.

The FEED design of the control centre was checked against ISO 11064 (1999–2013). To mention a few requirements:

1. Adequate space for its users, workplaces, and instrumentation. As a rule of thumb, each console position needs 25–50 m². In general, each operator position should also facilitate workspace for a second person during off-normal process conditions, for training purposes, and so on.

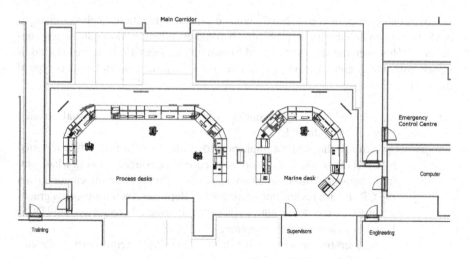

FIGURE 10.2 Front-End Engineering Design (FEED) layout design of the control centre.

2. CR access should be situated in front of the user(s). Thus, operators have visual control over who is entering/leaving the CR, and it is less disturbing than locations that are not in their field of view.
3. Operator positions having frequent communication with each other should be located close to each other and preferably within each other visual fields. This enables quick and reliable face-to-face communication.
4. Non-process-related communication shall not disturb operations in the CR. Staff who are not part of the operational 24/7 team may be tempted to use the CR for meeting purposes and acquiring information. Process operators need a quiet workplace to be able to concentrate on their cognitive tasks.
5. Operator workload should be on an acceptable level, not too much and not too little work. Operator workload may be estimated based on the number of controlled process variables, the expected rate of off-normal signals, communication with field operators, and thus the number of field operators, and communication with colleagues in the CR. The last variable can be related to the number of process flows between process units allocated to each operator position (Pikaar, 1992).

Control Centre Layout – Failure!

Both entrances to the CR as well as walking routes were at the operators' back. The corridor between CR and Emergency Control Centre (ECC) blocks quick and easy communication between both areas. Assuming the need for good communication in off-normal situations, direct visibility, and easy access is both recommended and best practice. We recommended removing both corridors to restructure routing and reduce operators' hindrances. Also, this would add some additional space to the CR and create direct communication lines between ECC and CR.

Based on our experience in ship design, we expected that these steelwork changes could be implemented without problems. It could even be done after steelwork construction. However, the shipyard refused to engineer changes, threatening large additional costs due to construction delays. The communication shown below is typical for this type of issue.

HF Professional: Did the project consider direct communication (visual contact) between ECC and CR?

Company Operations: that was not implemented due to layout constraints as reported by CONTRACTOR. It does require careful consideration as to what communication and information are required in the ECC for information from the CR. Please review and recommend. Replica screen for process graphics at ECC in order to not interrupt the distributed control system (DCS) Panel Operator may be an option.

Contractor: Consideration was given, but due to other design requirements, the current configuration was adopted.

Certainly, refusal to consider recommendations can happen. However, we were also expected to remove our recommendation from the ergonomic study report. If not, there might have been no report approval. I will come back to this "integrity issue" in the section on document control.

Control Room Layout – An Improvement!

FEED documents specified nine workstations, suggesting a maximum number of nine operators. The difference between four and nine work positions raised questions. Maybe, this had to do with the role of supervisors. Frequent face-to-face communication between supervisors and operators was expected. Also, the supervisor's office was rather small (21 m²) for three desks. Hence, we suggested creating supervisor workstations next to each CR console (Figure 10.3). Nonetheless, the role of the supervisors was never clarified to us.

Control Room Furniture – Success!

The initial CR layout (Figure 10.2) showed non-adjustable furniture without adequate legroom below the desks. The furniture was preordered by the ICSC at their tender for the project three years earlier! It was easy to convince the ICSC about this issue. However, this modular worktable system was part of the ICSC purchase order. They needed to go to the EPC with a change order request. Internal ICSC, EPC, and Company decisions took several months. Meanwhile, we developed a custom-built design, which was accepted in the end. A valuable tool to achieve success has been our 3D images of the CR design.

Work Environment Issues – Success or Failure?

As registered HF professionals we know our limitations. For example, we are competent to develop a lighting plan, but for other environmental topics, we may only check engineering results for compliance with guidelines and best practices. Reverberation time is an indicator of the acoustic quality of rooms. Guidelines for

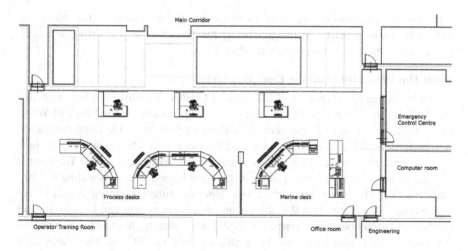

FIGURE 10.3 HF proposal for the control centre layout.

office environments aim for <0.50 seconds. Best practices for CRs suggest doing better than this, preferably <0.45 seconds. A reliable estimate can be made using Sabine's equation (Schomer & Swenso, 2002). During the review, we noticed in the noise philosophy document that a reverberation time >1.0 seconds was required for the CR. This might have been a typing error. So, we asked a question about this but got no reply.

The real problems started after we recommended additional noise-absorbing measures for the CR, based on an estimated reverberation time of 0.63 seconds. Apparently, engineering didn't like this recommendation. They used every trick in the book to delay progress, such as not responding within a reasonable time, postponing meeting dates, not showing up, asking irrelevant return questions about acoustics, and finally flatly refusing collaboration. Although frustrating, this was not uncommon, and therefore, relatively easy to accept. However, we were also urged to remove our recommendation from the ergonomic documents to get final approval.

Regarding the lighting plan, the above situation was repeated. The reason we developed competence on this topic is the utmost importance of glare-free lighting at multi-screen workstations. The shipyard provided a product catalogue of a local vendor. The catalogue included outdoor and navigation lighting, workshop lighting, bedside lamps, etc., but no suitable fixtures for offices or multi-screen workplaces. We explained what characteristics were needed and asked for relevant datasheets or data files for modelling a light plan. After several delays, we received the same catalogue three times.

Ultimately, approved by the ICSC, we developed a lighting plan based on the products of a marine lighting vendor from Norway. Obstruction at the yard became fiercer, including attempts to disapprove reports and payment milestones. What seemed to happen is that the shipyard was buying fixtures from a local vendor. They mounted the fixtures at the locations indicated in the HF-approved light plan, disregarding light source specifications. Glare issues would be expected, but it was impossible for us to verify. Although writing this chapter a few years after our activities for the Project, the vessel is not yet operational.

What Did We Learn from the Design Review?

According to our interpretation of the Code of Conduct for European Ergonomists,[1] we should report HF issues, even for ergonomic aspects outside the Scope of Work. In the Project, this was not appreciated by those responsible at the main contractor. It is understandable that engineers don't like comments on their work, but the lack of communication between HF and engineering was frustrating for us. We assume that more face-to-face communication might have helped, notwithstanding the language barrier. After several incidents, we stopped our time-consuming efforts trying to improve. It was also clear that the document control system can have an impact on the EPC behaviour. We learned we should never attempt to change an approved document. This is also frustrating but a characteristic of EPC-managed investment projects.

PROJECT PHASE 4: BITTERSWEET PHASE

After the review, a redesign of operator consoles and CR layout was undertaken. Amongst others, it resulted in a dedicated detailed design for the operator consoles. We introduced separate furniture for office tasks, printers, manual storage, and interior design (i.e. colours, materials, and finishes), including additional sound-absorbing materials. The results were presented in 3D images. ICSC and the Company were happy.

We recommended reducing the number of process screens on the consoles from typically 12 per work position down to 6–8. We had no doubt that this would be possible by optimizing the content of the process graphics (Pikaar, 2012). Our effort failed, but that was to be expected looking at CR solutions on the market with large arrays of tiled screens and many process graphics which, of course, determines the ICSC revenues. Of course, this topic was not in our scope.

For the marine console, the situation was different. Engineering responsibility was divided into four different disciplines: ICSC (processing units), communication, CCTV system, and radar. The total number of screens of different makes, sizes, and interactions was considerable. We counted 10 process screens, CCTV, and radar screens, two large annunciator panels, and separate instrument bays for maritime communication equipment. There was a long- and short-range radar and an expensive very high-resolution CCTV camera system for large distance recognition of approaching vessels. Both radar systems were to be installed as a backup for the CCTV in case of bad weather.

There was no clear idea about the operator tasks. Why was all this stuff needed since the vessel was anchored and would not navigate at all? Could one operator cope with all systems? From our point of view, it was not acceptable and possibly a safety risk. Initially, the EPC asked for our assistance in this matter. A purchase order

---

[1] https://eurerg.org/wp-content/uploads/2021/11/CREE-code-of-conduct.pdf

change glowed on the horizon. However, it appeared that our recommendations were not acceptable to the shipyard. All further communication ended abruptly.

What Did We Learn from the Bittersweet Phase?

A major part of an ergonomic study is the actual design of the CR and workplaces. This is a rewarding activity, always exploring possibilities to develop a unique and optimal result. However, HF is not a leading topic. Hence, you need to accept restrictions and compromises. Overall, we were happy with the result. Moreover, the ICSC was proud of the result. The bigger picture, however, is challenging. How to convince the stakeholders of the importance of, for example, a task analysis, an HF-inspired interaction (graphics) design, workload assessment, lighting, and acoustics? We believe that we could have done better regarding communication, provided the main contractor would allow or facilitate this.

Unfortunately, there is not much more to say about this phase. During the commissioning of the vessel and its process installations, HF is not present. If we are lucky, pictures appear on the internet about a successful naming ceremony or an official handover of the vessel to the Company. We could find pictures of the CR for the first project from the official opening ceremony. Organized operational feedback is seldom part of an HF Scope of Work. We tried many times to introduce this in the bidding, but generally, companies don't want to spend money on this. There is no doubt that feedback would be useful, and it is part of all theoretical design approaches. Pikaar and Caple (2021) offer recent insights into why case material is seldom published.

PROJECT PHASE 5 – DOCUMENT CONTROL

One could say that document control is a project internal way to fill in operational feedback. From an HF point of view, documents need to clarify (1) why specific solutions have been adopted and (2) what dependencies exist between workplace layouts, viewing distances to screens, readability, and the environment (e.g. light and ventilation). For these reasons, we try to integrate related factors into one report. Thus, a report on workplace design and CR layout could be one Ergonomic Study report. Another report addresses related environmental aspects (i.e. a focus on ceiling design).

However, in the first project, the scope of work specified 14 documents, including a document listing the documents that would be produced. Each document would require three submissions (1) for review, (2) for approval, and (3) final. Feedback from all stakeholders should be received within three weeks. In practice, comments came late. During the first project, documents also needed to be translated. Here, the ICSC took care of the translation, but we also learned that you need a certified translation agency, specialized in the terminology used in engineering and the particular type of industry.

For the Project, we were able to agree on seven deliverables, a 50% reduction compared to the first project; it looked like a success! These were:

1. Ergonomic Study report: A full description of the CR design.
2. Furniture descriptive notice: A detailed furniture specification, ready for the Invitation to Bid to furniture vendors.

3. Integrated ceiling design about lighting, ventilation, and noise control measures: It is usually issued separately, after approval of the CR layout and workplace design. Notably, this is a challenge because engineering requires this in an early project phase.
4. Wall and floor covering: This document could have been avoided by combining it with document 3.
5. CR Drawing 2D
6. CR Drawing 3D
7. Room Drawings: 2D drawings for each of other rooms in the Control Centre.

We believe that an ergonomic study report should be a "living" document, updated at each design step based on feedback from users. The idea of living documents implies informal feedback before formal submission. For documents No 5, 6, and 7, we intended to deliver the original 3D CAD file and sets of images. This idea did not fit in with the requirements of the document control department. The artefacts had to be written to undergo the document control process. Based on our experience from the first project, the Ergonomic study report also included an introductory statement (disclaimer): *"We would like to emphasize that at this phase of an HF contribution, we give our professional opinion. Our comments are based on generally accepted Human Factors knowledge, guidelines, and best practices. Our comments may not be in line with the current design. In those cases, please consider the comments open-mindedly"*.

The third project was managed differently and more proactively. The number of reports was reduced to four during discussions on the RfQ/Scope of Work. The reports and recommendations were appreciated. Document control was efficient, which can be attributed to the proactive EPC project management. In fact, here the living document approach came to life, resulting in only two formal revision cycles.

What Did We Learn about Document Control?

Document control is typical for this type of industry. Document controllers are powerful. However, the procedures do not easily allow for an integration of design aspects. In the Project, the recorded amount of work on document control was >40% of the total HF job. The procedures were time-consuming. Comments from different stakeholders did not become available at the same time and were not managed properly. Company comments were always late, and there was no effort to solve contradictory comments from different stakeholders. The Company and the EPC disagreed on several HF-related issues. However, the document control process didn't allow us to explain and discuss these issues. In practice, the three revision cycles became 2×3 cycles.

As mentioned earlier, integrity issues occurred, which had an impact on the document control–related work. In case HF recommendations were not implemented, we were expected to remove them completely from the report(s). Also, we were urged to describe the final design according to stakeholder decisions, as if it was an HF-approved result. It is difficult to tell whether this situation is typical for EPC-managed projects or had to do with the EPC lead and/or other persons in the engineering team. Notably, document control in the third project approximated 20% of our total work efforts.

DISCUSSION

"Why did they hire HF professionals, if there was no intention to change an initial design?"

ROLES

After reading this chapter, you might ask the question above. Clearly, our expectations on both the engineering process and HF content contributions didn't match with the ideas of the EPC. Partially, the mismatch has to do with different roles the HF consultant may have in large industrial projects. Without claiming scientific proof, I suggest the following different roles for HF consultants in large investment projects instead of cramming an ergonomic study in an EPC-managed project.

- *Company HF consultant, preferably before FEED*
  This role provides a powerful position. Our company has years of experience in this role, and you can read several of our case studies (Pikaar, 2007). Typically, this contribution takes place at and/or even before the FEED phase of a project. There is large freedom in design. The RfQ and/or Bid is based on mutual agreement on the HF activities needed.
- *HF review consultant*
  This role is about an independent validation and verification of an HF design developed by others. This activity is mandatory in Norwegian oil and gas projects (Johnsen et al., 2016; Pikaar et al., 2016) and can be mandated by Company rules. The consultant is contracted by or on behalf of the Company. The job is straightforward. Recommendations are taken seriously; it is a formal review! Costs can be calculated easily, and a fixed-price purchase order is no problem.
- *Detailed HF engineering*
  A lot of HF engineering must be done after the FEED phase when freedom in the design is limited. The HF consultant is part of a much larger team responsible for the final control centre implementation. Summarising the above experiences from the three projects, one needs to fight for results within the fixed budget.

Communication

At the time we received the RfQ for the first project, we didn't have much experience with the EPC and subcontracting settings. We were wrong in thinking that we knew the business. Communication, document control, a large multidisciplinary team, and all sorts of formalities took more time than expected. For the Project, we were aware of the risks involved but, nevertheless, our bid was in fact far too low. So, we won the bid but had to be careful about spending too much time, possibly unintentionally at the cost of communication. Communication was also hampered by travelling distances, different cultures, and language barriers.

Commitment

The success of a project also depends on the customers' understanding of HF. In cases of high customer HF standards, good results can be achieved. When HF knowledge is limited, achieving good results requires an additional effort for explaining HF's background and, of course, an interested audience.

Tightly related to customer HF standards is management commitment. If a company is purchasing the HF contribution, commitment, generally, is high, and generous funding is available. If the EPC or a subcontractor is responsible, commitment may be low as, *"We must do an ergonomic study because the company is asking for it (or we are obliged to apply ISO 11064). Please keep it simple and reduce costs"*. Hence, we learned two things. First, find out about commitment before starting the project, and second, accept more reluctantly suboptimal outcomes. We found this difficult.

HF Consultancies Are Small Enterprises

We are a small, specialized HF group. Actually, there are no large dedicated HF enterprises worldwide. If the Company and/or EPC has a low HF standard, only aiming for compliance with minimum requirements, there are two options. The first is to comply with their level of understanding and accept the risk of integrity issues. The second option is to reject the job. However, you may need the job!

According to my experience, the full scope and potential of HF are not always recognized in large investment projects. Hence, it would be helpful if the HF community could do more to promote the benefits and applications of HF. According to Pikaar and Caple (2021), projects missed substantial benefits because of a lack of commitment to HF.

LAST REMARKS

This case was about contributing to a control centre project during the detailed engineering phase. The impact of the HF contribution clearly depends on the project phase. There are better opportunities to integrate HF during earlier project phases. In detailed engineering, HF is only a small part of the job; you are a minor player. Others are not too happy to change their work or even to spend time getting to know something about HF.

In the Project, some typical events were probably more visible than usual. In some instances, integrity issues arose, such as the request to delete recommendations against professional responsibilities. This instance may not have been fully representative, but it shows the effects of some underlying processes, such as project leadership, communication, and the impact of document control systems.

Remember, HF is not about simply adding a nice touch, nor about the fulfilment of minimum legal requirements. HF is about the integration of the Human Factor in the design. This requires a commitment by project stakeholders, as well as good and extensive communication.

REFERENCES

Dul, J., Bruder, R., Buckle, P., Carayon, P., Falzon, P., Marras, W. S., Wilson, J. R., & van der Doelen, B. (2012). A strategy for human factors/ergonomics: developing the discipline and profession. *Ergonomics*, 55(4), 377–395. DOI: 10.1080/00140139.2012.661087.

https://www.sintef.no/globalassets/upload/teknologi_og_samfunn/sikkerhet-og-palitelighet/prosjekter/lyselng/criopreport.pdf

ISO 11064. (2000–2020). *Ergonomic Design of Control Centres – Multi-Part Standard.* Geneva, International Organization for Standardization.

Johnsen, S. O., Bjørkli, C., Steiro, T., Fartum, H., Haukenes, H., Ramberg, J., & Skriver, J. (2016). Criop: A Scenario Method for Crisis Intervention and Operability Analysis. Sintef Report A4312. Trondheim. Pikaar, R. N. (1992). Control room design and systems ergonomics. In: Kragt, H. (ed.). Enhancing Industrial Performance: Experiences of Integrating the Human Factor. Taylor & Francis, 145–164.

Pikaar, R. N. (2007). New challenges: Ergonomics in engineering projects. In: Pikaar R. N., Koningsveld, E. A. P., & Settels, P. J. M. (eds.). Meeting Diversity in Ergonomics. Amsterdam, Elsevier, 29–64.

Pikaar, R. N. (2012). HMI conventions for process control graphics. In: Proceedings of the 18th IEA World Congress on Ergonomics. Recife, Brazil. *Work*, 41(suppl 1), 2845–2852. DOI: 10.3233/WOR-2012-0533-2845.

Pikaar, R. N., & Caple, D. C. (2021). Challenges to engaging human factors/ergonomics practitioners to publish and present case studies. *IISE Transactions on Occupational Ergonomics and Human Factors*, 9(2), 67–71. DOI: 10.1080/24725838.2021.2006361.

Pikaar, R. N., de Groot, N., Mulder, E., & Landman, R. (2021). Cases of human factors engineering in oil & gas. In: Nancy L. Black, et al. (eds.), *Proceedings of the 21st Congress of the IEA, Volume III: Sector Based Ergonomics.* Springer Verlag, 42–49. DOI: 10.1007/978-3-030-74608-7_6.

Pikaar, R. N., DeGroot, N., Mulder, E., & Remijn, S. L. M. (2016). Human factors in control room design & effective operator participation. In: *Proceedings SPE Intelligent Energy International Conference and Exhibition.* Aberdeen, Society of Petroleum Engineers (SPE).

Schomer, P. D., & Swenson, G. W. (2002). Electroacoustics. In: Middleton, W. M., & Van Valkenburg, M. E. (eds.), *Reference Data for Engineers* (9th ed.). Newnes. DOI: 10.1016/B978-075067291-7/50042-X.

11 Deciphering the Knowledge Used by Frontline Workers in Abnormal Situations

Christopher M. Lilburne and Maureen E. Hassall
University of Queensland

CONTENTS

In the oil and gas industry and many other high-hazard work domains, frontline workers are tasked with detecting and addressing abnormal process deviations to maintain safe and efficient operations. These tasks are critical since, when not properly addressed, abnormal situations can rapidly escalate into catastrophic events such as the BP Texas City Refinery explosion which killed 15 workers (U.S. Chemical Safety and Hazard Investigation Board, 2007), the Tesoro Anacortes Refinery explosions which killed seven workers (U.S. Chemical Safety and Hazard Investigation Board, 2014), and the Soma mine explosion in 2014 in which 301 workers lost their lives (Düzgün & Leveson, 2018). Major accidents are still occurring and reoccurring across high-hazard industry sectors resulting in lost lives, significant injuries and/or illnesses, environmental harm, and asset damage. Additionally, there are reputational, legal, and social issues for individuals, corporations, and governments.

This topic is of much interest to us, who hail from the high-hazard industries. Chris worked in the oil refinery industry for nearly 10 years and has had roles as a process and process safety engineer prior to working in Human Factors and Ergonomics (HFE). Maureen has worked in the process, mining, and manufacturing industries for many decades. We both have much experience responding to incidents both operationally and as investigators and have delivered projects to attempt to address reoccurring incidents; we are passionate about eliminating work-related fatalities and catastrophes. From our experiences when incidents occur, including near misses, incidents, or major catastrophes such as those listed above, investigators often uncover gaps or mismatches in workers' knowledge. These "knowledge gaps" are commonly associated with a breakdown in the ability to diagnose the system state and/or respond in the manner required to mitigate disastrous outcomes. Thus, recommendations from investigations are often framed in terms of improvements to knowledge transfer, capture, application, and storage. Prominent examples come from the United States and the Netherlands. While they are extracts from very large investigations, their knowledge-based findings are enlightening.

For example, as part of the investigation into the BP Texas City explosion, the U.S. Chemical Safety and Hazard Investigation Board (2007) identified that *"the operator training program was inadequate..."* (p. 23) and recommended BP *"Improve the operator training program ..."* (p. 215). When investigating the 2014 Shell Moerdijk reactor explosions, the Dutch Safety Board (2015) found that *"The Panel Operators and the Production Team Leader were supposed to use their knowledge and experience of this start-up process to adjust the gas and liquid flows, as needed. However, they lacked this experience"* (p. 37). More recently, when investigating the DeRidder Paper Mill explosion, the U.S. Chemical Safety and Hazard Investigation Board (2018) found that *"numerous operators demonstrated a general lack of knowledge about the tank, its role in operations, or hazards that it posed"* (p. 77) and recommended the company *"provide workers with periodic training to ensure they have an understanding of all process safety hazards applicable to..."* (p. 89). There are many other similar examples publicly available in investigation reports.

The quotes, which are not intended as critiques, provide some detail as to what knowledge was lacking. However, what are missing are specifics of *what* knowledge is required to successfully manage a similar abnormal situation in the future and how this knowledge should be best conveyed to workers so it will be retained and applied when required. From our experience in leadership and engineering roles, we have found that implementing recommendations that tend on the side of generality can be frustrating as they do not convey the specific knowledge that needs to be intuitive or available for quick reference by those responsible for hazardous systems or processes. Depending on the investigation techniques employed, it is possible that more specific recommendations cannot be generated, often making it hard for a typical plant manager, engineer, investigator, designer, or operator to define what "good" looks like.

Driven by our passion to eliminate fatality and catastrophic events, we set out to discover what good might look like. We investigated occasions where frontline operators have successfully arrested or managed a sequence of events before significant harm occurred. Our study focussed on successful outcomes as we believed it might be useful to understand the various conditions and preconditions that support personnel to identify, diagnose, and respond to abnormal situations. We thought that capturing such behaviour may lead to insights that allow more effective responses beyond revealing knowledge gaps after adverse events. The "human as hero" has been addressed previously by Reason (2008) and others who describe how people can use their knowledge and experience to adapt and improvise to successfully manage complex systems and situations (e.g. Jamieson & Vicente, 2001; Rasmussen, 1974;

Rasmussen, Pejtersen, & Goodstein, 1994; Woods, Dekker, Cook, Johannesen, & Sarter, 2010; Woods & Hollnagel, 2006).

Several seminal case studies also highlight where frontline personnel have successfully prevented disasters by identifying and solving novel problems in real time. One example is the US Airways Flight 1549 that was ditched in the Hudson River in 2009 (New York, U.S.). In this instance, *"the professionalism of the flight crew members and their excellent CRM [Crew Resources Management] during the accident sequence contributed to their ability to maintain control of the airplane, configure it to the extent possible under the circumstances, and fly an approach that increased the survivability of the impact"* (National Transport Safety Board, 2010, p. 91). Similarly, during the Apollo 13 event, disaster was *"averted only by outstanding performance on the part of the crew and the ground control team which supported them"* (National Aeronautics Space Administration, 1970, p. ii).

Our hypothesis that understanding success is important when responding to adversities is further supported by the work of Rasmussen et al. (1994) and Hollnagel (2014) who assert that success comes from ensuring that things go right rather than solely focusing on preventing failure. In short, we sought to answer the question "What knowledge do frontline workers need to successfully manage high-risk activities?" by seeking to answer the question, "What knowledge do frontline workers *use* to successfully control high-risk work?"

To answer our question, we launched an HFE research project to explore the decision-making and actions of frontline workers in an operating oil refinery. Specific organisation was chosen as Chris had previously worked in the industry which meant that experience with the technology and processes as well as professional connections within the company help to gain access to operators. If followed that Chris was able to facilitate the interview discussions and he spoke the "language" and understood the context and significance of the events being discussed by the operators. This allowed Chris to ask questions needed to gain an understanding of the knowledge that underpinned how workers successfully resolved real-world, abnormal situations that otherwise might have been catastrophic. To help elicit detailed information about operator decision-making, we chose to structure the interviews using the Critical Decision Method.

The Critical Decision Method is a semi-structured interview strategy designed to elicit detailed information from decision-makers retrospectively about non-routine events (Hoffman, Crandall, & Shadbolt, 1998; Klein, Calderwood, & MacGregor, 1989). The Critical Decision Method follows several standard "sweeps":

1. Identify a suitable event
2. Interviewee describes the event uninterrupted
3. Interviewer identifies and probes one or more key decision points (Where? Why? How? etc.)
4. Hypothetical questions are asked. For this exercise, two standard questions were:
   a. How would a novice have performed?
   b. What would help in a similar or identical future situation?

Chris conducted these interviews with oil refinery operators over the course of about four months to cover different operational shifts in each trip. Operators were provided with an information pack prior to the visit. About half a day was dedicated to each visit and the individual interviews took between 30 minutes and 1 hour each. Preparation involved reading other Critical Decision Method interview research and conducting practice interviews with colleagues who were not involved in this work. Following each refinery visit, there was a debriefing between us, the authors of this chapter. Overall, ten volunteer refinery operators were interviewed, and we analysed the data related to 11 discrete events.

The debriefing process proved beneficial to both of us. It allowed Chris to review and reflect on the process, the event, and the insights that were being elicited. It also allowed Maureen to understand the range of events being discussed and the level of detail being captured. It proved to be a useful check to make sure that the approach being used was delivering the responses and information that could be further analysed to answer the question "What knowledge do frontline workers *use* to successfully control high-risk work?"

Through a specific case below we describe the analysis tools used, including the original and modified versions of the Decision Ladder template and a newly developed tool called DeciMap.

PREVENTING THE WORST

An oil refinery field operator with more than 20 years' experience was nearing the end of his 12-hour shift. He was walking back to the central control room following final field checks when he noticed that the seal of a running pump had failed. The pump was pumping ~270°C hydrocarbon and the leak had begun forming a flammable vapour cloud, which represented an extreme hazard. The operator's significant knowledge of the plant meant that he was able to instantaneously identify the leaking pump, the material leaking, and the associated hazards. Critically, the operator also knew that this situation was unfolding approximately 15 m away from a fired heater. The presence of the heater was a significant ignition source that could detonate the vapour cloud. Around this time, a second field operator also attended to the situation. As the second operator was not called by the first operation, he most likely attended by chance or because he could hear the commotion from nearby. At some point during the initial diagnosis, one of the attending operators radioed into the central control room to report the situation.

Immediately, the two operators worked together to isolate the leak. Between them, they started the spare parallel pump, turned off the running/leaking pump, and closed its inlet, outlet, and other associated valves to isolate the leak. Alternatively, they would have used fire hoses to suppress the leak, but this option was rejected due to time constraints, proximity of the leak to ignition sources, and the expanding size of the already large vapour cloud. The selection of the response strategy appears to have been an intuitive rather than conscious trade-off between the options to isolate the leak, suppress the leak, or evacuate. The operator stated that if the fired heater wasn't so close, they would have been more likely to use fire hoses to manage the leak. Using fire hoses is an inherently safer strategy to resolve the leak as it reduces personal exposure to fumes and reduces the potential for a fire to break out.

Even after the two operators completed the pump changeover and valve isolation, the vapour cloud was still forming, meaning that the leak had not been effectively isolated. The operator knew that this would be due to a pipe still being open to the pump, meaning that they must have missed something. Therefore, the operators scanned the pipework for an open connection. Although access to the pipework connected to the leaking pump was not an issue, it was hard to see due to the leak. Ultimately, after less than a minute, an open pump warmup line was identified. The line was isolated, the leak stopped immediately, and the gas cloud quickly dissipated.

The event lasted for five to ten minutes before the leak was resolved and the plane returned to a safe state. Expressing the impact of his experience on the event, the operator interviewed said that he was able to bring the situation to a safe conclusion because of having seen similar and analogous events in the past. Specifically, he had seen the same or similar events *quite a few* times on other pumps at the facility and had executed a similar response two or three times.

THE ANALYSIS JOURNEY

TRADITIONAL DECISION LADDER ANALYSES

The Decision Ladder template, which was developed by Rasmussen (Rasmussen, 1974, 1986; Rasmussen et al., 1994) and later described in detail by Vicente (1999), is one of the standard Cognitive Work Analysis templates used for decision analysis. Its flexibility and adaptability made it a logical place to start this exploratory work. It can be used to document "the different modes of perceptions and processing" used by decision-makers (Rasmussen, 1974, p. 26). The sequence is presented as "the steps a novice must necessarily take to carry out the sub-task" (Rasmussen, 1974, p. 26). Broadly, the template describes three phases of decision-making namely, situation analysis (left leg), knowledge-based reasoning and planning (top part), and execution of tasks (right leg) (Figure 11.1).

However, although a decision could be made by a novice operator in this order, it has been shown by several authors that decision-making typically does not follow this sequence. A person with experience could, for instance, have the knowledge and experience to shortcut the novice's approach, especially when dealing with familiar, low-risk tasks. These shortcuts are manifested as associative leaps between states of knowledge ("leaps") or shunting between data-processing activities and states of knowledge ("shunts") (Rasmussen, 1980). In our case, the operator made observations about the condition of the plant, gas cloud, location, etc. and then made a cognitive "leap" to an understanding state of the plant and hazards. This was driven by the operator's years of local experience; he did not need to stop to think about it thoroughly. Similarly, he didn't consciously identify the steps required to swap the pumps and isolate the leak. This is a shunt from identifying the "chosen option" to allocating the tasks to isolate the leak.

In general, we found that using the traditional Decision Ladder approach allowed for a description of the patterns of operator decision-making. During the analysis,

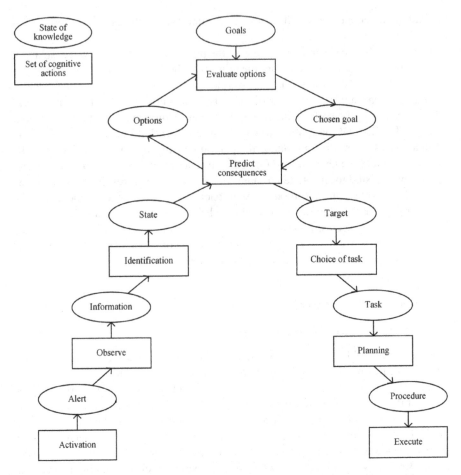

FIGURE 11.1 A generic decision ladder template. (Adapted from Vicente, 1999 and Rasmussen, 1974.)

many examples of shunts and leaps described by the refinery operators were reported and documented, leading to a long and rich catalogue of decision-making shortcuts (i.e. shunts and leaps). However, we realised that it was difficult to explicitly document some details of the operators' decision-making. Specifically, we observed that operators are often faced with competing goals and ambiguous or changing situations which can result in the generation and comparison of different response options.

For instance, in this case, the operator alluded to weighing up the competing goals of personal and plant safety and continuity of operations. The operator described how the team implicitly, rather than explicitly, selected their response plan. In addition to intuitively generating and selecting an option, there were also responses generated and reviewed using explicit knowledge (e.g. pre-established procedures, plant drawings, and own or peer's experiential knowledge). For this reason, we modified the Decision Ladder template, as explained below.

Modified Decision Ladder Template

As a result of the challenges experienced when using the traditional Decision Ladder template, we attempted a second round of Decision Ladder template analysis using a modified version (Lilburne & Hassall, 2019). The major difference between the traditional and modified Decision Ladder template is that it shows a more explicit description of option generation, comparison, and selection (Figure 11.2). The differences are:

- Additional states and actions have been added relating to understanding "gaps and ambiguity" followed by option generation, modification, and evaluation.
- The cognitive action of "predicting consequences" is now annotated in three stages: "project future states", "awareness of potential future state(s)" and "assessing for acceptability against goals".

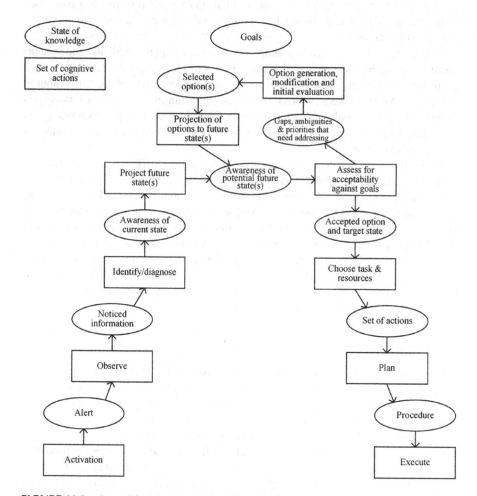

FIGURE 11.2   A modified decision ladder. (From Lilburne & Hassall, 2019.)

- "Goal" is decoupled from the decision process where often it is linked to "Evaluate".

By using this modified template across all cases, we collected, where reported by the refinery operators, what future states they were considering and how these states aligned with or violated their goals at different time points. It allowed us to map out and distinguish between the different states of knowledge and cognitive actions as described by the field operators involved, especially regarding strategy selection. Thus, the modified template made it possible to catalogue how decision-makers address any ambiguities or unknowns they confront during abnormal operations. For example, in the case described above, the operators' goals were to maintain a forward flow while resolving the leak and maintaining the safety of personnel present. However, they did not report going through a process of option generation. They intuitively "knew" which response they were going to attempt. In other cases, a violation of goals resulted in a conscious weighing up of response options (e.g. commence an emergency shutdown versus continue operations).

The modified Decision Ladder template, however, could not explicitly and systematically capture how knowledge was used. For example, it did not allow for detailed capturing of experience with past events, use of critical real-time data, or the input of colleagues. All these were reported to play an important role across various aspects of decision-making and response. Workers reported using their prior knowledge to know how to respond, where to look for more data, or when and where to seek out advice and expertise.

Overall, it was clear from the interview material that there was a rich volume of data relating to the knowledge underpinning the refinery operator's decision-making. The challenge we faced was how to best document and understand the data. One option was to annotate the completed Decision Ladder templates to explain the various shunts and leaps described by the operators. While this would be a relatively straightforward exercise for one or two cases, we thought that it was not scalable to the whole set of cases. Additionally, descriptive annotations would not be sufficiently systematic, meaning that we would risk losing consistency, explainability, and repeatability of the analyses. Therefore, while the initial analysis rounds provided a necessary baseline, further work was needed.

DECIMAP

A search of available HFE approaches did not reveal a tool, method, or template that suited our purposes to identify the knowledge used by frontline workers when successfully responding to abnormal events in different decision-making contexts and relying on different combinations or their own or their peers' experiential knowledge and a range of explicit sources of information and knowledge. Therefore, we developed our own method to identify and link the different types of knowledge used during different parts of the decision process. The tool, called DeciMap (a shortening of *deci*sion *map*ping), is more thoroughly documented in Lilburne (2021). It is a swim lane method where a decision process is converted to a timeline to which different types of knowledge are mapped. Our aim was to develop a graphical, easy-to-use, reliable, and useful process.

The concept for DeciMap came about from teaching Rasmussen's AcciMap (Rasmussen, 1997; Rasmussen & Svedung, 2000) to a cohort of Chemical Engineering students. We realised that a key feature of the AcciMap aligned with our objectives is to *graphically* map things (e.g. knowledge and experience) at different levels of abstraction to an event timeline. While DeciMap does not resemble the AcciMap, it was the starting point that led to adopting the swim lane concept.

We then commenced working on determining how the "lanes" should be defined through several iterations, applications to different case studies, independent comparisons, and verifications. We also engaged a third person familiar with the AcciMap process and our project to review and critique our development work. The third person added much value and acted as a relatively independent challenger who ensured that our work was more purposeful and deliberate rather than just *stuff*.

In the final concept, there are six swim lanes or rows (Figure 11.3). The upper rows are the *Knowledge* rows and the lower are the *Event* rows. The knowledge rows were drawn from an extensive review of knowledge, management, and human factors literature as well as from the coding analysis of the "Why?" and "How?" questions asked during the Critical Decision Method interviews (Lilburne, 2021). The rows capture the different sources of knowledge reportedly used by the decision-maker to carry out cognitive reasoning. There are three sources of knowledge, namely explicit knowledge and implicit knowledge gained by first- or secondhand experience.

The event rows document the actions of a decision-making, their evaluation and judgement processes, and how the actions or knowledge of others interacted with the

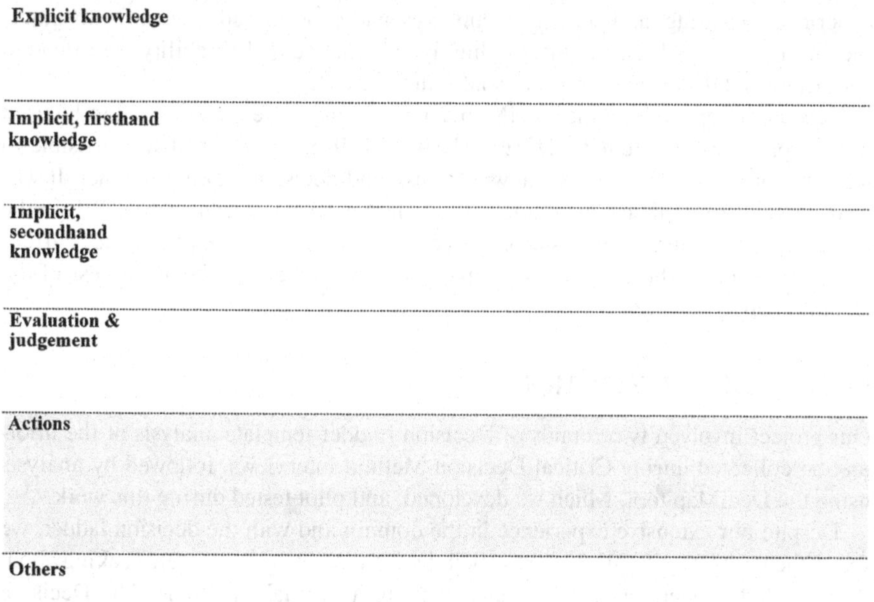

Explicit knowledge

Implicit, firsthand knowledge

Implicit, secondhand knowledge

Evaluation & judgement

Actions

Others

FIGURE 11.3 A blank "DeciMap" analysis template. (From Lilburne, 2021.)

TABLE 11.1

Definitions Used in the Initial DeciMap Analysis

Knowledge rows	**Explicit knowledge**	This category includes explicit knowledge used in decision-making, such as observations, system read-outs, and drawings
	Implicit, firsthand knowledge	This includes knowledge gained through firsthand experiences
	Implicit, secondhand knowledge	This includes secondhand knowledge gained by hearing about or witnessing events (e.g. "war stories" from colleagues)
	Evaluation and judgement	Describes the evaluation and judgement aspects of decision-making
Event rows	**Actions**	Describes the actions of the decision-maker. For example, "Making a radio call" or "Changing a control setpoint"
	Others	How people other than the decision-maker impacted a decision-making process at the time of the event. For example, "Advice given by a colleague"

Source: Adapted from Lilburne (2021).

decision-making. These rows were created based on a pragmatic way to display an event and decision sequence and align with what can be typically captured during a Decision Ladder template analysis. Definitions for each of the rows are shown in Table 11.1.

There were two outputs of the DeciMap exercise. First, we developed a catalogue of different types of knowledge sources that can be mapped to various aspects of decision-making. In the case described above, these were relatively simple to document. However, other operators described more complex interplays between their experience, training, and specific events they had experienced. Some, for instance, recalled decades-old plant failures as highly informative in their ability to respond to analogous and different events much later in their career.

Second, it appears that the DeciMap approach may have the potential to be used more broadly as a systematic HFE analysis tool. In this project, it was possible to identify some patterns of how knowledge use and decision-making interact. It was also possible to compare operators with different levels of experience. While the impact of these initial analyses is hard to assess, the results were promising. Further work now needs to be done to test, refine, validate, and assess the usefulness of the DeciMap approach.

CONCLUDING REFLECTIONS

Our project involved two rounds of Decision Ladder template analysis of the information collected during Critical Decision Method interviews, followed by analysis using the DeciMap tool, which we developed, and pilot tested during this work.

Despite our extensive experience in the domain and with the decision ladder, we found the template difficult to use to elicit the detailed insights we were seeking about the knowledge operators used when managing abnormal situations. The Decision Ladder template was useful but didn't go far enough. Specifically, we struggled with using the decision ladder to help elicit and represent implicit and explicit knowledge used by operators. This frustration led us to develop and test the DeciMap. We were familiar with using the AcciMap and were delighted that we found that a similar approach could prompt and visually represent in an easily digestible form the knowledge used by operators.

Overall, the initial applications of DeciMap using oil refinery operator interviews seemed to be effective and allowed for a more thorough examination of knowledge used in decision-making. We managed to gain initial insights into how operators use different types of explicit knowledge as well as first- and secondhand implicit knowledge to avert disasters. By describing different types of knowledge used by operators using these categories, we discovered some nuances that excited us because we believe that they could be used to improve the support given to workers to help them manage abnormal situations. For example, we found detailed examples of how recent abnormal events are usefully recalled alongside distant (sometimes decades old) abnormal events to determine response plans.

In performing this work, we learned a lot of lessons that can be related to any HFE work so we would like to share. The first lesson we want to share is the value of having domain expertise when interviewing workers. We certainly found value in being able to enter a workplace and "talk to the talk". It allowed Chris to develop a good rapport with the operators and draw out detailed incident accounts.

We also learned to watch out for the self-selection and analysis bias. Self-selection bias is when individuals select themselves into a group (e.g. interviewees for case studies), which produces a biased sample leading to skewed and not representative results. This form of bias is difficult to address but needs to be considered when identifying trends or findings. The analysis bias regards the inevitable subjectivity of the interviewer and analysts collecting and processing information. In this project, we each performed the different analyses independently and then compared and discussed the results to help minimise analysis bias. We found that this process also helped us identify areas where we had differences in our interpretations.

Another lesson that stood out for us regarded the strengths that can come from the combination of multiple HFE approaches. This was unintended at the start of our project, but, on reflection, it was advantageous. Often, within the HFE literature, there is some priority put on understanding which tool is the most accurate, insightful, or usable in different contexts. While there is an element of competitiveness, literature shows the different strengths and limitations of different approaches. In our study, the combination of the original Decision Ladder template, our modified version of the latter and the DeciMap we created was more insightful than if each had been considered individually.

While multiple tools proved beneficial, beginning the work with a relatively simple approach was also useful. This was Chris' first significant HFE project; when we started the research, he was well-experienced in the oil refining sector but had minimal HFE experience. Chris found the Decision Ladder particularly of value as the first HFE tool to use, as the Decision Ladder template is highly accessible, flexible (*viz.* forgiving), well published and documented, and quick to learn. This also meant that it was a straightforward method to *pitch* to a prospective industry partner, clear concept, mostly in plain language with existing examples of value-adding work.

Other more complex HFE approaches are less widely published and therefore harder for inexperienced individuals to use effectively.

Finally, making modifications to existing HFE templates and experimenting with creating our own approach also aided this work. Modifying the Decision Ladder template acted as a forcing function requiring us to do a deep dive into the origins of the template, review how others have used it in the intervening decades and give serious consideration to which version of the template should be applied for this work. An alternative would have been to select a template with little thought, populate it, and move on.

Similarly, developing DeciMap was a useful exercise requiring exploration of how decisions can be documented and how different types of knowledge can be conceptually, practically, and philosophically categorised. This, again, continues a tradition of HFE practitioners developing their own approaches, tools, and methods. It is perhaps a big leap for a novice HFE practitioner to begin developing their own tools and methods. However, on reflection, there are plenty of published methods, tools, templates, and variations thereof, each designed and tested for its own bespoke purpose. For this work, we could find no existing tool that would meet our analysis goals. An alternative to DeciMap would have been to simply extract and code different knowledge types into a spreadsheet or other tool, leading to a much less rich outcome. Clearly, if DeciMap and the Decision Ladder template modifications are to have an impact beyond this work, more testing and application are needed. However, from an analysis and HFE practitioner perspective, we believe that the time invested in understanding, developing, and documenting operational knowledge was extremely worthwhile. We hope that by publishing our work here, others will try it to produce interesting findings and the nuances in knowledge use that underpins both successful and unsuccessful attempts to manage safety critical abnormal situations in the industry. And we also hope that these findings will lead to work system design innovations that better support workers to produce successful outcomes, thereby preventing fatalities and catastrophes.

REFERENCES

Dutch Safety Board. (2015). *Explosions MSPO2 Shell Moerdijk (English version)*. Retrieved from www.safetyboard.nl

Düzgün, H. S., & Leveson, N. (2018). Analysis of soma mine disaster using causal analysis based on systems theory (CAST). *Safety Science, 110*, 37–57. doi:https://doi.org/10.1016/j.ssci.2018.07.028

Hoffman, R. R., Crandall, B., & Shadbolt, N. (1998). Use of the critical decision method to elicit expert knowledge: A case study in the methodology of cognitive task analysis. *Human Factors: The Journal of Human Factors and Ergonomics Society, 40*(2), 254–276. doi:10.1518/001872098779480442

Hollnagel, E. (2014). *Safety-I and Safety-II: The Past and Future of Safety Management*. Farnham, United Kingdom: Taylor & Francis Group.

Jamieson, G. A., & Vicente, K. J. (2001). Ecological interface design for petrochemical applications: supporting operator adaptation, continuous learning, and distributed, collaborative work. *Computers & Chemical Engineering, 25*(7–8), 1055–1074. doi:https://doi.org/10.1016/S0098--1354(01)00678--0

Klein, G. A., Calderwood, R., & MacGregor, D. (1989). Critical decision method for eliciting knowledge. *IEEE Transactions on Systems, Man, and Cybernetics, 19*(3), 462–472. doi:10.1109/21.31053

Lilburne, C. M. (2021). *Knowledge Use in Safety-Critical Decision Making.* (PhD Thesis). The University of Queensland.,

Lilburne, C. M., & Hassall, M. E. (2019). Modifications to the Decision Ladder to match frontline workers' critical decision making. *Proceedings of the Human Factors and Ergonomics Society Annual Meeting, 63*(1), 347–351. doi:10.1177/1071181319631329

National Aeronautics Space Administration. (1970). *Report of Apollo 13 Review Board.* Washington, DC.: National Aeronautics and Space Administration.

National Transport Safety Board. (2010). Loss of Thrust in Both Engines After Encountering a Flock of Birds and Subsequent Ditching on the Hudson River US Airways Flight 1549, Airbus A320-214, N106US, Weehawken, New Jersey, January 15, 2009. (NTSB/AAR-10/03). Washington DC.

Rasmussen, J. (1974). The human data processor as a system component. Bits and pieces of a model (Technical Report No. Risø-M-1722). Danish Atomic Energy Commission Risø, Roskilde, Denmark.

Rasmussen, J. (1980). The human as a systems component. In H. T. Smith & T. R. G. Green (Eds.), *Human Interaction with Computers.* London: Academic Press.

Rasmussen, J. (1986). *Information Processing and Human-Machine Interaction: An Approach to Cognitive Engineering.*: New York, NY: North-Holland.

Rasmussen, J. (1997). Risk management in a dynamic society: A modelling problem. *Safety Science, 27*(2), 183–213. doi:10.1016/S0925-7535(97)00052-0

Rasmussen, J., & Svedung, I. (2000). *Proactive Risk Management in a Dynamic Society.* Sweden: Swedish Rescue Services Agency.

Rasmussen, J., Pejtersen, A. M., & Goodstein, L. P. (1994). *Cognitive Systems Engineering.*: New York, NY: Wiley.

Reason, J. T. (2008). *The Human Contribution: Unsafe Acts, Accidents and Heroic Recoveries.* Boca Raton, FL: CRC Press LLC.

U.S. Chemical Safety and Hazard Investigation Board. (2007). *Investigation Report - Refinery Explosion and Fire.* Retrieved from https://www.csb.gov/file.aspx?DocumentId=5596

U.S. Chemical Safety and Hazard Investigation Board. (2014). *Investigation Report: Catastrophpic Rupture of Heat Exchanger (Seven Fatalities)* (2010-08-I-WA). Retrieved from https://www.csb.gov/file.aspx?DocumentId=5851

U.S. Chemical Safety and Hazard Investigation Board. (2018). *Non-Condensable Gas System Explosion at PCA DeRidder Paper Mill.* Retrieved from https://www.csb.gov/file.aspx?DocumentId=6052

Vicente, K. J. (1999). *Cognitive Work Analysis: Toward Safe, Productive, and Healthy Computer-Based Work.* Mahwah, NJ: Lawrence Erlbaum Associates.

Woods, D. D., & Hollnagel, E. (2006). *Joint Cognitive Systems: Patterns in Cognitive Systems Engineering.* Boca Raton, FL: CRC/Taylor & Francis.

Woods, D. D., Dekker, S., Cook, R., Johannesen, L. J., & Sarter, N. B. (2010). *Behind Human Error* (2nd ed.). London: CRC Press.

12 The Tyranny of Misusing Documented Rules and Procedures

Nektarios Karanikas
Queensland University of Technology

CONTENTS

As with my contribution to the previous book in the same series, which was about safety insights (Karanikas, 2020), the context of my cases is the highly demanding operational environment of a military aircraft squadron. While writing these lines, I realised that the two years when I was a chief engineer with an enormous pile of responsibilities for staff and equipment left valuable marks on me. Although before and after this role I had several important duties at operational, tactical, and strategic levels, I believe my successes and failures during my duties as a chief engineer were more impactful on my growth as a professional. This relates to the position of this role in the organisational structure. On the one hand, you are responsible for the health, safety, well-being, and professional currency of the staff trusted in you and for the availability and airworthiness of the aircraft and other ground equipment in the squadron. On the other hand, you are accountable to the squadron commander, and, by extension, the operational base commander, and the rest of the management levels. Simply put, you cannot hide from anyone. What I was doing was visible in all directions, upstream and downstream.

As with most organisations, especially large ones, rules and procedures were inextricable parts of our work design. We could find them everywhere, and they drove the "who, where, when, what and how" of our activities. Compliance with those documents was expected from everyone and was ingrained in us since the early days of our military and engineering education and training. Rules and procedures were the reference points to service aircraft, schedule and perform maintenance, license technicians, keep records, submit reports, etc. Any deviation from those, especially the ones that dictated our safe interactions with aircraft (e.g. refuelling or inspections) could incur implications for staff, supervisors, and managers, including myself, regardless of whether the outcome of a procedural deviation was an incident/ accident or not. Of course, when the result was adverse (e.g. injuries or damages), the personal consequences across any organisational level involved were worse.

I was supposed to monitor everything and ensure compliance. You could wonder how this was possible with tens of staff at different places performing various activities within diverse environments and varying operational demands. I was wondering too but worries alone can change nothing. I had to do my job and help everyone keep safe, productive and happy. However, sometimes someone needs to be a sort of rebel and challenge established rules and procedures when these do not serve the needs of the staff and the work. The cases that I am sharing focus exactly on two of my attempts to change the status quo. However, before moving to the stories, the next section includes my understanding and reviews some literature about rules and procedures. I believe this can help you interpret my decisions and actions described later in the cases I share.

RULES AND PROCEDURES AS PARTS OF WORK DESIGN

I presume when someone mentions procedures, our minds often bring up pictures of short or long documents prescribing what we are supposed to do in specific circumstances, how to use our new gadget, how to enter and analyse data in a software package, how to maintain a machine, how to drive our car, etc. When referring to rules, our brains typically recall the written and unwritten "laws" of what others expect us to do and achieve or not. We should not interrupt others when speaking, we are expected to help others in need, we should report an accident to the authorities or a medical condition to our insurer, etc. Simply put, rules usually represent the "what" and procedures complement rules by describing the "who, where, when, and how" of matters. The only interrogative question missing from the list above is the "why". I will come back to this important aspect later in the chapter.

Nonetheless, from a more abstract level, procedures as stepped approaches to complete activities within a set of rules (i.e. boundaries of what we can do) are present in the very early years of our lives. They are not written, but this does not render them inexistent. Our brains observe, process, sometimes imitate what others do and achieve and, with the coordination of our body parts, use procedural steps to carry out daily activities to yield the intended outcomes as dictated by internalised rules. Our internal software continuously grows, changes, and adapts to what we sense, and it continually develops new and modified "if, then, else" loops to complete several missions underpinned by these various stored rules, which also can change with time.

However, in the work context, even if we are given rules and we memorise these, we do not always have the luxury to become shadows of more experienced professionals and learn through practice how they do the job successfully, presumably according to the rules. Some tasks that we could be asked to perform in the future might not be possible for others to demonstrate if an operational need does not arise. Even when such a demonstration becomes possible physically (e.g. on-the-job training on mock equipment) or virtually for a limited time (e.g. in simulated environments or video recordings), nobody can claim that we really master tasks after a couple of successful attempts following demonstration and performance under supervision.

Furthermore, experienced staff might not be the perfect examples of how work should be expectedly done according to procedures and rules. Long and rich work experience increases our confidence, overconfidence might lead to overestimating our capabilities, and this, in turn, leads to drifts from procedures without prior consideration of what could go unintentionally wrong. Indeed, someone with substantial work experience might deal successfully with the unexpected results of procedural deviations, but an inexperienced employee might just freeze, run away, lose control of the situation, etc., with dire consequences. That was one of my major concerns in my role as a chief engineer several years ago. I was worried when novice technicians were just imitating experienced workers who did not follow procedures exactly, but the former individuals did not yet have developed capabilities to confront unfamiliar situations.

Add to the above the fact that experienced workers with valuable knowledge might leave an organisation and our brains can forget procedural steps and rules, especially when we do not practise them often. We might also omit steps, usually when we do something very often. All these are the main reasons for documented rules and procedures. As Pasquini and Pozzi (2005) explain, different procedures aim mainly to facilitate the integration between operators and equipment, including, of course, the cooperation among operators. However, the critical aspect when designing procedures is that we must view the human-tool-procedure system as a whole within their physical and social work environment (Pasquini & Pozzi, 2005).

For instance, whereas Bleyl and Heller (2008) observed that Standard Operating Procedures in healthcare can increase workload but support high-quality patient care, in the same industry sector, Thomas and Spain (2012) warned that safety is more than policies, processes, and procedures. Amongst other work stressors, the authors above referred to the nature of the work performed, comprehensibility and accessibility of procedures, management styles, supervisory capacity, teamwork, and collegial relationships. Referring to rules, Hale and Borys (2013b) identified two approaches in the literature, namely (1) top-down enforcement of rules as static, comprehensive limits of freedom of choice and (2) bottom-up constructivism where rules are dynamically, locally, and situationally constructed through competence and ability to adapt them to different work realities. As a reconciliation of those two approaches, Hale and Borys (2013a) suggested a framework of rule management that acknowledges the necessity of monitoring and adapting rules through worker participation and regular and explicit conversations between users, supervisors, and technical, safety, and legal experts. The combinations of top-down and bottom-up paths when creating objectives and procedures were also claimed as indications of the balance between systematic management and systems thinking in an organisation whereby performance variability is controlled and accepted to different extents (Karanikas et al., 2020).

Indeed, in the mining maintenance context, Kanse et al. (2018) identified that procedure management strategies played a more positive role when providing learning opportunities than punishment for deviations, and users were more keen on complying when engaged in procedures' design and review. Bringing the above to the aviation context, where both my success and failure cases refer, Carim et al. (2016) proposed that procedures should not be viewed as controlling constraints but resources to support actions because situations encountered by pilots can be far more complicated than what procedures are designed for. Therefore, amongst other strategies for the design of procedures, these should provide pilots with choices rather than mandatory steps. As Provan and Rae (2020) concluded, although we need rules and procedures within organisations to enable work to be performed safely, the challenges of the dynamically changing work environment warrant a critical review of the applicability of rules and procedures within an organisation.

SAME PROCEDURE, DIFFERENT ACTORS

One of the critical preflight phases is what we call the "last-chance" check. In our organisation, we performed this just before the aircraft entered the runway to line up for take-off. This check includes a visual inspection of several critical systems to detect any issues that could threaten flight safety (e.g. impacts on tyres after taxiing towards the runway, leakage from hydraulic and oil pumps and accumulators, or loose screws on surface panels). If the technician finds a problem, depending on its magnitude, he/she might call for a more experienced technician to check again or can immediately instruct the pilot to cancel the operation and return to the squadron.

The "last-chance" tasks are performed based on a prescribed procedure (i.e. what and how to inspect and in what order) in a very noisy environment because of the aircraft engines running. Also, the technician must move below the aircraft surfaces, meaning that there is no eye contact with the pilot for some of the activities' duration. To avoid any unintentional harm to staff by moving surfaces (e.g. wing flaps), the pilot lifts his/her arms high and away from any lever, switch, etc. when the technician asks for it. The pilot moves the aircraft surfaces only when in eye contact with the technician upon the instructions of the latter. Those surface movements allow the technician to detect any anomalies (e.g. fluid leakages or inconsistent surface movements). When the task is finished without findings, the technician signals to the pilot, he/she steps away from the aircraft, and the mission continues.

In this process, there are two additional risk control measures in place. First, as in most cases of servicing and maintenance tasks with the engines running, there is a fire extinguisher close to the inspection area. In case of a fire in the engine or other aircraft systems (e.g. leakage of hydraulic on the very hot braking system), the technician can prevent or mitigate the expansion of the fire by using the extinguisher while also communicating with the pilot through standardised visual signals. The second control measure is an extra technician who keeps a safe distance from the aircraft under inspection but continuously maintains eye contact with the pilots. This ensures that if something goes wrong with the technician inspecting the aircraft (e.g. medical condition) when not visible by the pilot, the supervising technician will inform the flight crew and instruct them further (e.g. shut down the engines).

Notably, the "last-chance" supervisor usually monitors two inspections at the same time by keeping a distance that allows an adequate view angle to cover both aircraft, which are usually stopped next to each other with a safe distance between them. If there were more aircraft to be inspected, they either had to wait for the inspections of the first pair of aircraft to finish or we had an extra team available to conduct the "last-chance" tasks. Moreover, although the inspections could be carried out by any fully licenced mechanical technician, the procedures foresaw that we should appoint only an experienced or inspector mechanical technician to the supervisory role. The classification for all technicians across the organisation was Level 3 for fully licenced, Level 5 for experienced technicians, and Level 7 for inspectors.

All the above were great, and everyone complied to ensure the safety of everyone and everything. However, just after a few weeks in my role, I noticed that sometimes, especially during night operations, there was much urgency and tension when it was time for the "last chance". Sometimes I was hearing this on my portable radio device and other times I was witnessing it with my own eyes; teams of technicians running around frantically to get into the vehicle that would drive them to the "last-chance" area. Anything that has to do with anxiety and speed in safety–critical environments, and at work in general, is concerning to me. I had to understand what was going on.

It took me only one discussion with the staff to realise the reasons for the situation above. Simply put, there was low system capacity because of staff shortages. There were only a few Level 5 and 7 mechanical technicians per shift, about 3–4 during day operations and only 1–2 in the night shifts. These specialists had to attend several works at the flight line and the maintenance hangar to ensure that enough aircraft were available on time. When it was time for the "last chance", it could happen that some of these specialists were occupied with critical duties that did not afford them the luxury to stop and restart their jobs or they were away from the squadron for other tasks (e.g. transferring components to and from the depot). Thus, although the maintenance scheduling office was trying its best to have experienced technicians available during flight operations, there was usually only one Level 5 or 7 mechanical technician who could undertake the supervisory role by rushing from one task to this one, together with the rest of the technical team.

I first investigated whether I could increase Level 5 and 7 staff capacity day and night. Unfortunately, this was not possible for several reasons (e.g. lack of experienced Level 3 technicians to be elevated to Level 5, absences because of personal leave or needs for operations at remote locations, necessary resting period between shifts). Then, I looked at the work itself. I started by consulting with the documents, including the organisational directive that established the "last-chance" procedure a few years later. The intent was commendable as it ensured operational safety; no technician denied the need to have this procedure. However, I realised that, while issuing the procedure, nobody had considered the resource implications and the particularities of each squadron. It was a document that applied everywhere uniformly, regardless of the context. Whereas in several cases the "last-chance" area was next to the squadron, in our case, the teams had to drive about 5 km to reach that area. It could not be different because of the design of the runway in our base and the need to perform these inspections just before the aircraft were about to line up for take-off. Hence, I could not change this part of the work system.

Afterwards, I witnessed twice the task conducted in real time. The technicians complied with the procedure, and no step seemed to me unnecessary. However, the second time I observed the task, I had my eureka moment! Yes, I concurred that a Level 5 or 7 technician must have the supervising role, but why a mechanical technician only? I did not notice any supervisory task which needed mechanical expertise! Monitoring the area for abnormalities, helping with extinguishing the fire if any existed, and signalling the pilot were tasks that all experienced technicians were licensed and able to do regardless of specialty.

The next day, I invited all Level 5 and 7 technicians to my office, including mechanical, electrical, and avionics and fly-by-wire specialists. I shared with them my observations and thoughts above. The mechanical specialists were happy but sceptical about whether the rule could change. The rest were unhappy because they would be required to do more. The efficiency driver of human survival struck! Nonetheless, I explained to them that the workload would not be significantly higher because it would be shared amongst all Level 5 and 7 personnel, and the task mostly necessitated monitoring from a safe distance. I also explained the benefits for their mechanical expert colleagues and everyone in the squadron. Nobody amongst us would like to hear about or experience an injury because of the rush to squiz people and tasks and drive fast to and from the "last-chance" area. It did not take long. Collegiality and solidarity prevailed.

The next thing to do was to receive approval for this change from management. I discussed our idea with the squadron commander, and he agreed I would submit a formal request to the base management. I did so by describing the particularities of our squadron (i.e. far from the "last-chance" area), the staff capacity limitations, the gains from this recommended change to allow any specialty for the supervisor role, the absence of any additional risks, and, most importantly, that no extra training or other resources were necessary. We just needed a formal change of the rule so that we could amend our local procedures.

However, having experienced severe response delays in the past or hesitancy to change rules, this time I added something extra to my request. Instead of just asking for approving the change, I informed the base management that we would start implementing the revised procedure in two weeks if we had not received an answer prohibiting the change. For the military context, that was rebellious! You are only supposed to request and wait, possibly nudge in the meantime. Until you receive a formal response, you must follow the top-down imposed rules and procedures. Of course, it did not take long before I received a phone call from the base's maintenance director. He shouted, "Who do you think you are? How dare you arbitrarily change procedures decided at senior organisational levels?" With all due respect, I asked back "Do you deny that we have a problem to solve? Did you find anything in the recommended change that is not workable or inflicts additional risks? Does the base have any alternative way to help me solve this issue?" He hung up.

Indeed, the base never formally replied to my request, positively or negatively. Back then, they informed me orally that they were seeking approval from the tactical level because the directive had been issued by them and affected several bases and their nested squadrons. Unsurprising development, I said to myself, for a large, highly structured, and bureaucratic organisation. Nonetheless, in the squadron, we proceeded as planned. We changed our internal procedure and since then everything was running more smoothly and with less stress. The mission of work redesign to match the local context was accomplished. If I am not wrong, the tactical-level rule has not changed since then.

WITH THE BOOK OR BY THE BOOK?

In the defence sector, the mantra "prepare for war during peace times" drives everything. Indeed, this is what this sector does when not in warfare. One of the critical activities is to run drills that simulate war conditions to evaluate the system's performance and improve. In our organisation, such drills were taking place a few times throughout the year at all operational bases. They were announced by the level responsible for evaluating our readiness. These would be the operational, tactical, or strategic levels. I cannot even count how many times I was involved in those drills. I admit that they were great experiences. Everyone was alert and prepared to contribute to defending against the virtual enemy! They were also valuable opportunities to discover any shortcomings and perform better next time at individual, team and organisational levels. After each drill, a report from the inspecting authority shared the findings, both positive and negative. The latter ones concerned, operational bases and squadrons should come up with actions to rectify any issues. We were supposed to recommend how to change our work systems and/or design!

Notably, those drills were not focussing only on whether we could deliver the right service/product on time, but also on whether we could deliver it safely and sustainably. No defence organisation wants to suffer from lower warfare capacity and capabilities due to staff injured and aircraft damage because of internal issues. We needed staff and equipment available for and being capable of continuous operations to secure a win without harming citizens and destroying the natural environment! Therefore, the assessors were looking at several aspects beyond the success of aircraft operational missions, including physical and cyber security, health and safety, training and licencing, maintenance capabilities, impacts on communities and the environment, etc. Documented rules, procedures, and checklists about where and what to check were inextricable parts of these evaluations.

During one of these drills organised by the tactical level, I was sitting together with an assessor to discuss several aspects. Those discussions were important for the assessors to gain a better understanding of the context and ask questions about their findings (e.g. remedies we already planned for but were possibly delayed). Also, these conversations were opportunities for us to share with other organisational levels our struggles and ideas. One technician came close and apologised for interrupting us. He seemed nervous. The technician reported that one assessor rebuked a team performing a "turnaround" because most of them were not holding their checklists. Moreover, the assessor informed them that he would report this as an important finding. During the drills, as we knew we are evaluated, we felt much pressure. We should perform perfectly!

The "turnaround" process includes aircraft inspection, servicing with fuel and oil, and reloading with weapons in a very limited time, assuming that the aircraft just returned from a battle and should be ready the soonest possible for the next mission. The same process happens during training missions but in a different setting. The "peacetime procedure" expects that a Level 3/5 mechanical technician will inspect the aircraft, replenish oils and hydraulics if necessary, and refuel it with the support of another licenced staff member. Then, the armament team (2–3 technicians of all levels) will load the aircraft with the weapons required, and a Level 7 mechanical technician will perform the last inspection.

Notably, there are specific works not permitted on the aircraft when loading weapons or refuelling. This is to avoid catastrophic consequences, such as the unintentional activation of weapons (e.g. someone touches the wrong switch). Next, everyone signs off the maintenance forms, and the aircraft is reported available for its training mission. Of course, if the mechanical or armament technicians notice any problem, further actions are taken. After all these steps, the pilot checks the forms and inspects the aircraft together with a technician, and, if everything looks good, the rest of the mission continues. The servicing part by the mechanics took about 40–45 minutes, and the duration of the armament depended on the weapons needed. The latter could add 30 minutes for the "simplest" configuration to several hours for the "heavy" configuration. For a simple configuration, add the time for signing off forms and the rest of the steps outlined above, and we get about 90 minutes for the entire process.

The difference in wartimes (and drills!) is that the 90-minute process must be completed in less than 30 minutes with all technicians working simultaneously to prepare the aircraft for its mission. To achieve this on time and safely, the mechanic team is usually supported by another technician, and the Level 7 mechanic undertakes the role of the "turnaround" supervisor. Since different tasks are carried out in parallel, the procedure has been designed in a way that avoids overlaps and bottlenecks (e.g. staff performing a different job in the same area) and ensures that safety risks are controlled (e.g. no electrical checks when weapons are loaded or during refuelling). Therefore, imagine a team of six persons plus the refueller amidst several equipment, tools, and weapons working hard to complete their operation- and safety–critical jobs in a very limited time with a restricted area defined by the limited dimensions of the military aircraft. The adrenaline gets to the top levels!

Well, the good news is that we were all gradually trained for this before any drills. Nobody was suddenly asked to perform this highly demanding job without prior experience. As part of this training, we were consulting with the "turnaround" checklist to understand the differences from the regular process and we were accompanied by experienced persons to perform the tasks in the specific order by holding our checklists, then gradually working with other technicians to complete the process, and, finally, performing the "turnaround" faster and faster until we lowered its duration to the desired length. We were not rushing things further as we were aware that this would increase health, safety, and operational risks because the process was already designed with almost no slacks and no room for deviations.

Nonetheless, as we were gaining more experience in our tasks during the "turnaround" training, we left aside the checklists because they became obstacles rather than supportive items. Imagine running around to complete your job the soonest possible with safety and quality while your one hand is occupied by the documented checklist. Impossible! However, the supervising technician was always holding the checklist as a reminder of the order and type of tasks to be carried out by the different team members at each time point. When the supervisor noticed something abnormal, she/he provided appropriate instructions or reminders (e.g. next task to be completed). After the process was finished, the supervisors gave us feedback and explained to us any slips and lapses by referring to the checklist. Each time, we studied the checklist again, and we did it better the next time. Therefore, whereas we required inexperienced trainees to hold the checklist ("with the book"), we expected experienced staff to work "by the book" while the supervisor monitored everyone while holding the checklist and consulting with it.

Unfortunately, that was not the perspective of the assessor who criticised the technicians in the specific case. I invited the assessor and his team leader to discuss this. The assessor repeated that he noticed exactly what I described above; the technicians performed the "turnaround" while only the supervisor held the checklist. This meant that, from my point of view, there was no gap between what we had planned and what the team was doing, or between Work As Imagined and Work as Done as Hollnagel (2014) defines it. To verify this further, I asked the assessor whether he observed any safety and quality issues with the process and whether all steps were completed in the right order and on time. He admitted that the "turnaround" he attended was exemplary. Great news!

Then I asked him why he persisted on this issue of all technicians holding the "turnaround" checklist. He replied that his own inspection checklist required so. I kindly asked to read his checklist, which mentioned that the process should be carried out *according to* the approved "turnaround" checklist. Nowhere did I read that the technical team should hold their checklists while doing their job. Obviously, it was a matter of different interpretations. The assessor insisted that his checklist required a "with-the-book" process, and I was advocating that the checklist meant "by-the-book" performance of the tasks. When it came to arguments to defend our positions, the assessor had nothing else to use apart from his understanding of the checklist and the fact that he had applied the same concept during other drills, and everyone complied. Well, this last thing was concerning to me.

I explained to him that the risks of holding a book during the "turnaround" process with the expectation to consult with it (otherwise, there would be no need to hold it) was dangerous. The staff focussed on completing their critical tasks in a limited time while not inflicting delays, harming others, or damaging the aircraft and their equipment. Had they had to read their checklist before each step, or sets of steps, it would be highly distracting. Not only would this delay the whole process, but it would also divert the cognitive capacity of technicians to an unnecessary activity. We were indeed prepared very well for the drill – warlike environment by consulting the checklist – and now it was time to demonstrate what we can do! Can you imagine someone holding the driving manual of his/her car or the book with the road signs and reading it while on the road? The assessor's interpretation of his inspection checklist made no sense to me. He provided no convincing answer to my "why?"

The lead assessor was more political. He admitted that he agreed with my interpretation and our practice, but he could not enforce his opinion on the other assessor who, on the one hand, also shared my concerns, but, on the other hand, stuck with his decision to report his observation as a finding. Every additional attempt from my side to deter him went wasted. It was not only about our squadron or whether we would be "charged" with a noncompliance finding. It would not be the first time. I was more worried that he would impose his interpretation of using checklists on other bases and squadrons. Within military structures, rarely would someone challenge the opinion of senior management levels. In this case, armed with my knowledge of human factors and safety, I dared to present substantive arguments, and I openly refused to comply. Would others do the same? Would they resist this insensible perspective?

After a few weeks, we received the report, and the alleged finding was there as expected. The squadron commander asked me about it. I described to him all the above. He was silent for a minute or so. Then, he said, "I trust you will do the best for our staff". I replied, "I will not comply". We decided to report back to senior management on this finding by stating "continuous efforts to comply". That was a typical close-off statement throughout the whole organisation for items we did not have something tangible to suggest or a good excuse. We never put in any extra effort to comply with the "with-the-book" requirement during "turnarounds", at least while I was the chief engineer in the squadron. We continued to do our best to deliver effective on-the-job training and minimise health, safety, and operational risks.

Could I have done something differently? Back then, I felt any official letter dedicated to this misinterpretation of how checklists are supposed to be used would get lost in the bureaucratic maze of our large organisation. In hindsight, I could have at least tried. I regret it now. I could have also invited the assessor to a drill performed in his way, meaning "with the book" and not just "by the book". However, I did not want to impose this unnecessary experimental risk on staff to reinvent science for the sake of someone's denial to accept it. The best I could do, and I did, was to continue to be close to the technicians, consult with them, share with them my knowledge, experience, and advice, and work out together the best ways to minimise risks.

A FEW LAST THOUGHTS

Whether written rules and procedures can be friends or enemies depends on everyone's contribution. Authors, auditors, operational managers, and users must all work in tandem and genuinely design and redesign documented ways of work. This of course lies more on the initiatives of senior leadership and management to create bridges with others and create a psychologically safe environment where everyone can respectfully discuss and debate. I believe that, in this way, even the term "Work as Imagined" would be obsolete because it implies that others decide about our work on our behalf and without us.

Instead, a collaborative and open environment would lead to create Work as Agreed, which should be continually revised based on shop-floor experiences of Work as Possible. This would establish the "why" of matters to supplement the "what" of rules and the "who, where, when, and how" of procedures. Work as Imagined can be as dangerous as Work as Done when the former does not match the work context and when locally emerging practices are not informed by collective discussions about and assessments of side and unintentional risks.

Furthermore, I argue that not all agreed works have to be documented to be enacted. The work environment is a living environment, and as life cannot be put on a paper, the same applies to work systems. Indeed, a legitimate question would be "How much should we document?" You can call me theoretical and idealist, but I do not think that one size fits all when it comes to the length and number of procedures. Inexperienced workers might need detailed information about both rules and procedures. Experienced workers who have internalised rules and procedures might need summaries or more abstract representations of rules and procedures, while also having the opportunity to get back to the detailed ones. Here, certainly, the importance of awareness and collegial culture play a hugely important role. We must be able to alert each other to get back to the detailed procedures when our practices threaten the integrity of the system and continually risk-assess together and feedback to the system on any necessary deviations.

Therefore, assuming a mature organisational culture, my response to you would be "as much as it gives each worker cohort the opportunity to access the information needed to perform the job as expected". If you decide to have lengthy prescription documents for everyone, I understand. We do not have the resources to tailor everything to everyone. Also, maybe you do not trust the workers and vice versa to establish a Work-as-Agreed and Work-as-Possible environment. Trust is another crucial organisational parameter, but its discussion can be long, and it is outside the scope of this chapter. Nonetheless, in this case of a highly prescribed work environment, we could also describe what we expect from each cohort when it comes to compliance. Do we want tasks to be performed by the book or with the book and when and by whom? Do we think that holding the book will make people safer? Do we expect people to apply our procedures religiously or adapt them to their context based on the approach recommended above? What are we prepared to do if such adaptations usually succeed but sometimes fail? I urge us all to think again.

REFERENCES

Bleyl, J. U., & Heller, A. R. (2008). Standard operating procedures und OP-Management zur Steigerung der Patientensicherheit und der Effizienz von Prozessabläufen. *Wiener Medizinische Wochenschrift*, *158*(21), 595–602. https://doi.org/10.1007/s10354-008-0607-y

Carim, G. C., Saurin, T. A., Havinga, J., Rae, A., Dekker, S. W. A., & Henriqson, É. (2016). Using a procedure doesn't mean following it: A cognitive systems approach to how a cockpit manages emergencies. *Safety Science*, *89*, 147–157. https://doi.org/10.1016/j.ssci.2016.06.008

Hale, A., & Borys, D. (2013a). Working to rule or working safely? Part 2: The management of safety rules and procedures. *Safety Science*, *55*, 222–231. https://doi.org/10.1016/j.ssci.2012.05.013

Hale, A., & Borys, D. (2013b). Working to rule, or working safely? Part 1: A state of the art review. *Safety Science*, *55*, 207–221. https://doi.org/10.1016/j.ssci.2012.05.011

Hollnagel, E. (2014). *Safety-I and Safety-II: The Past and Future of Safety Management*. Ashgate.

Kanse, L., Parkes, K., Hodkiewicz, M., Hu, X., & Griffin, M. (2018). Are you sure you want me to follow this? A study of procedure management, user perceptions and compliance behaviour. *Safety Science*, *101*, 19–32. https://doi.org/10.1016/j.ssci.2017.08.003

Karanikas, N. (2020). Necessary Incompliance and Safety-threatening Collegiality. In N. Karanikas & M. M. Chatzimichailidou (Eds.), *Safety Insights: Success & Failure Stories of Practitioners* (pp. 139–148). Routledge. https://doi.org/10.4324/9781003010777-13

Karanikas, N., Popovich, A., Steele, S., Horswill, N., Laddrak, V., & Roberts, T. (2020). Symbiotic types of systems thinking with systematic management in occupational health & safety. *Safety Science*, *128*. https://doi.org/10.1016/j.ssci.2020.104752

Pasquini, A., & Pozzi, S. (2005). Evaluation of air traffic management procedures—safety assessment in an experimental environment. *Reliability Engineering & System Safety*, *89*(1), 105–117. https://doi.org/10.1016/j.ress.2004.08.009

Provan, D., & Rae, A. (2020). Rules and Procedures. In *The Core Body of Knowledge for Generalist OHS Professionals* (2nd ed.). Australian Institute of Health & Safety. https://www.ohsbok.org.au/wp-content/uploads/2020/04/12.3.1-Rules-and-procedures.pdf

Thomas, S., & Spain, J. (2012). A Manager's Perspective: Patient safety/employee safety: Much more than policies, processes and procedures. It's a culture! *Canadian Journal of Medical Laboratory Science*, *74*(1), 16–18.

13 Creating Ownership and Dealing with Design and Work System Flaws

Stasinos Karampatsos
Hellenic Air Force

CONTENTS

The context of both cases below is the everyday operations of an aviation organisation with state-of-the-art, high-performance fixed-wing and rotorcrafts, piloted or unmanned. The organisation operates its aircraft daily and with a high frequency, and it can conduct operations anywhere in the world with proper maintenance capability and support for operations. Considering the high procurement costs of planes and the dire consequences of an in-flight failure (e.g., human injuries or losses, equipment damage, or a crash into populated areas), the requirements for preventive aircraft maintenance and repairs are treated with attention to detail and extreme seriousness (EU, 2018). Work of this kind comprises most tasks of an aircraft maintenance organisation, supported by the logistics chain. Of course, apart from safety, the availability of serviceable aircraft is paramount for the missions of business units.

This specific organisation has a quality management structure that conforms to international standards and is based on the general principles and guidelines of Deming's Plan-Do-Study-Act philosophy (Evans & Lindsay, 2008).[1] There is dedicated support for logistics and airworthiness and engineering departments in accordance with internationally accepted directives (EU, 2014; EDA, 2021). Each on-site maintenance unit includes flight-line services (i.e., operational/basic-level maintenance) and base maintenance (i.e., intermediate-/second-level maintenance) but no depot maintenance capabilities (USDS, 2021). Organisational elements that deal with quality assurance and health and safety issues are also included in this structure. The following cases are from my observation as a maintenance unit manager, having undergone approved training as an aircraft engineer, being a subject matter expert in

---

[1] https://deming.org/explore/pdsa/

DOI: 10.1201/9781003349976-13

quality and risk management, being a certified project manager, and having extensive experience in managing medium to large teams.

TASK OWNERSHIP IN PROCESS IMPROVEMENT

System faults can obviously hinder the safe use of equipment and/or the execution of organisational missions. Hence, a good deal of maintenance organisation's efforts goes towards remedying such faults. However, repairs are usually the smaller part of the work than the volume and duration of the tasks of preventive maintenance, where standard practices for the design and use of aircraft apply. Preventive maintenance aims to counter two major deterioration categories. One of those is wear and tear due to the operation of aircraft. For example, the accumulation of a specific amount of flight hours requires the replacement of certain components that are subject to stress during flights. If not replaced, those parts could fail due to cracks or other fatigue-related issues. The second category regards wear due to environmental factors and the interaction of aircraft systems with their surroundings (e.g., oxygen or ultraviolet radiation). Such maintenance takes place at regular time intervals, usually regardless of the operation frequency of aircraft.

Aircraft manufacturers primarily utilise steel, aluminium, and other nonferrous alloys to achieve reduced weight and higher performance (FAA, 2018). However, in the environments that the aircraft operate, these structural materials are prone to corrosion, especially in combination with high mechanical stress during flight operations. Consequently, dealing with corrosion issues is a primary concern for aviation organisations and a major part of preventive maintenance activities. In this case, corrosion issues arose in critical structural elements of the aircraft. Failure/damage critical components are the ones that support loads and if damaged or degraded in strength can lead to structural failures. These failures could result in serious or fatal injuries, or extended periods of operation with reduced safety margins (FAA, 2021). The corrosion prevention schedule is the process through which aviation organisations deal with such problems.

In this case, the type and extent of the corrosion damage found far exceeded the allowable limits. Thus, it was necessary to alter the corrosion prevention schedule of aircraft, apply corrective measures, and carry out all these as soon as possible to avoid potential in-flight mishaps. The business unit dealing with basic-level aircraft maintenance had the responsibility for dealing with the transition from the "business as usual" state to a totally new one, within a relatively short timeframe; initial steps had to be taken within a week, and a complete overhaul of the schedule should be completed in under a month. This would involve a drastically differentiated maintenance schedule and increased maintenance requirements. However, the strict timeframe did not allow for implementing broad organisational-level changes (e.g., completely new roles and responsibilities for each maintenance level) or allocating considerably more resources (e.g., committing additional technicians, time, and equipment already necessary for other activities).

Thus, the top-down-imposed goal – I had in my role as the maintenance unit manager – was clear: incorporate the new maintenance requirements in the ongoing work without affecting operations, as soon as possible. Dealing with such a situation falls within the definition of a project, where project and change management principles apply (PMI, 2021). Having a clear scope and schedule, I set forth to create an action plan for the necessary changes. The plan was to unfold in two different directions. One was to foster ownership of the changes throughout all organisational levels, both upstream and downstream; the importance of such ownership will become clear below. The other was to incorporate these changes into the already existing organisational processes through process improvements and avoid any sudden shift in the operational focus. The main mission of the business unit was primarily to support flight operations, and the resources available were allocated in such a way as to prioritise that. Therefore, the plan required all stakeholders, from front-line workers to senior management, to be informed and involved accordingly. Also, for the plan to succeed, the changes required had to be clearly defined. These changes could be organisational, task-related, procedural, or related to roles and responsibilities throughout the organisation or only the business unit. It was the burden of the business unit and mine to design these in accordance with the new maintenance requirements, implement those within my purview, and propose and support their acceptance with the minimum possible resistance from the organisation.

The new maintenance schedule was to be implemented as soon as possible. While the plan was initially put in motion in the business unit, the corrosion prevention tasks were assigned to the base maintenance level. The reason was that its staff was more competent and could perform the necessary work faster and with higher quality according to the specific instructions issued by the airworthiness and engineering departments of the organisation. However, it soon became clear that the implementation of the new maintenance schedule was creating undue confusion and disruption in the everyday work schedule.

Different teams were assigned unrelated tasks in the same workspace and were tripping each other up. The sequence of work created work duplication and delays. For instance, functional tests of the aircraft were scheduled right after it was freshly painted, or the aircraft was to be washed right after lubrication; such cases were creating a need for rework (e.g., repaint or re-lubricate). The base maintenance supervisor was very clearly distraught at the mess of this situation and was fervently advocating for a clear separation of tasks and reallocation of human resources. Having dedicated teams dealing with the tasks in a linear procession would be a way to solve rework issues. However, it would also mean having workforce allocation restrictions that would greatly reduce the capacity of the business unit to use its resources adaptively to fulfil requirements for other maintenance works.

As the business unit manager, taking the complaints of the staff seriously, I assembled a team of engineers and master technicians to record and assess the work area setup, the organisational structure chart, available resources, and work schedule data. At first, the solution obvious to the team and me was to separate work areas and workflows. Concurrent maintenance tasks and corrosion prevention tasks were to be avoided; otherwise, there was a high risk for mistakes and/or omissions. Therefore, I initiated a study by recording the team's findings and suggestions to determine the best way forward. We recorded the tasks that needed to be included in the new corrosion prevention schedule as a separate, detailed, and independent workflow. Charts were created where the inputs of every step in the work were analysed: the number of staff required, standard work times, sequence of works, necessary equipment, work area required, facilities and amenities, safety measures, and data management.

After the team designed a new corrosion prevention schedule implementation process, a model run was performed to collect data and create a benchmark for the process. After comparing these data with the resources available, it became clear what changes were necessary for the business unit procedures and workspace allocation. A special focus was also given to the type and amount of training the staff would have to undergo. Any shortage of equipment, inadequate safety measures, and/or insufficient facilities were also addressed. Overall, the approach aligned with Deming's philosophy of Plan-Do-Study-Act (Evans & Lindsay, 2008) and Juran's quality improvement process (Juran & Godfrey, 1998) and used best-practice quality management tools and techniques (e.g., process control, Pareto charts, and cause–effect diagrams).

Importantly, informal interviews with the staff, after every step in the process, and brainstorming meetings with in-house subject matter experts played a decisive role in the study my team performed. I formed the team, appointed a lead, facilitated and coached them, and coordinated their activities to record and resolve the users' requirements. I must clarify that *what* needed to be achieved was set in stone by the airworthiness and engineering departments; we did not have any authority and space to challenge this. Hence, we could only figure out by whom, where, and how that would happen, and make it so with the optimal use of the available resources. As typical for any cost-minded organisation, the changes had to be implemented with the minimum investment in human resources, equipment, and facilities.

The whole business unit took an active part in the study, contributing to data collection and putting forward suggestions and recommendations. The respective conclusions and suggestions were forwarded up the organisation's management ladder. In addition to the detailed report, to pass along the complete image of the situation and ensure a good understanding by the stakeholders, a campaign took place to inform senior management about the problems that had arisen and the possible solutions. The campaign included on-site surveys and visits by the senior management, including persons with the authority to allocate additional funds and resources. Notably, such campaigns were outside the usual modus operandi of the organisation, and because they required involving all the rungs of the organizational ladder, they were reserved for special cases only.

During the campaign, one by one, upstream management realised the importance of having a clear and precise view of the problem at hand and offered their support. Admittedly, the organisation management culture of ownership played an important role, making my job easier. On the other hand, it also required presenting the case for the changes in a concise and persuasive form that could be carried along from rung to rung to the final decision-makers. However, when the top brass saw the problem themselves, it was much easier to understand the solutions offered. Sometimes, one must put their finger into the print of the nails to believe.

Within 6 months of the start of the campaign, the organisation implemented the suggested changes in their totality. These involved separate stages where each organisational level, starting at ours (i.e., the business unit), applied changes within their authority. A separate work subsystem was created to deal with the corrosion issue, assigning additional resources, providing training where requested, and allocating dedicated facilities and equipment. This led to the successful resolution of the issue at hand, with minimal delays or impact on ongoing operations, and minimised or eliminated possible hazards.

Following the new system of work, statistics on findings and work performed were systematically collected, analysed, and fed back to the airworthiness and engineering departments. This way it was possible to verify and validate the successful resolution of the issue, both regarding the technical issues and the process improvement. The implementation was gradual and involved the inclusion of both external and business unit resources, adjusting the responses and updating the process policies along the way.

The primary lesson from this case is that the active involvement of all stakeholders in the chain of operations was the key to the successful resolution of the problem. The front-line stakeholders are often those with the clearest and most precise view of the issues faced. Their intuition and perception can offer the required solution, and this co-ownership of the solution facilitates its implementation. Also, the provision of accurate information about the problem and suggested solutions to those with the authority to make changes is essential; it should be undertaken in a way that these persons also take ownership of the need for changes.

Admittedly, an organisation's culture and structure are quite often difficult to steer in a different direction, and organisational size correlates with increased difficulty. Moreover, sweeping changes imposed by external actors are rarely met with open arms and broad acceptance. Creating a culture of participation throughout the organisation has multiple returns and is a prerequisite for success.

UNDETECTED DESIGN AND WORK SYSTEM FLAWS

In contrast to the case above, there are instances where the actors might fail to anticipate and prevent the escalation of a problem. Omissions or oversights during the initial design phase can create hazards downstream. These may take quite some time to surface, whether in the system or a process. Within the same organisational context described in the case above, refuelling activities were part of our daily routine in the business unit supporting flight operations and servicing aircraft. Work of this kind presents inherent hazards due to the nature of the materials handled. For this reason, staff performing such tasks is specifically trained for handling these hazards, and there are stringent safety measures in place (USAF, 2008).

It is a standard practice in the civilian (FAA, 2021) and military (USAF, 2008) sectors to use fuel pressure systems, especially where a high volume of fuel is moved. The fuel hoses used in aviation are manufactured to specific standards (USDD, 2011) and submitted to rigorous testing to ensure safety during their use. Also, the operation and maintenance of aircraft fuelling equipment are usually strictly controlled, with the respective procedures codified in checklists and, sometimes, in legislation (NYC, 2014). These procedures assume a standard mode of operations and specific aircraft, trucks, and hoses (FAA, 1974). By mode of operations, we mean the frequency and volume of fuelling component use, both daily and for the duration of its service life. Increased usage can, and usually does, lead to increased maintenance requirements. Edge cases where the system is within design scope, however marginally so, can reveal hazards that should have been considered in the initial design of the system and associated procedures, but were not.

In our case, flight operations were of high frequency, and the turnaround time was quite short; each time, the aircraft should be refuelled as soon as possible. As such, the daily number of fuelling operations was high. The aircraft were serviced by a small number of fuelling trucks, operated by specially trained staff. The latter worked according to established procedures to unreel the hoses, take safety precautions, fuel the aircraft, and reel and stow the hoses. The design characteristics of the trucks required manual operation of the reeling drum since there was no motor provided to power the reel; this significantly increased the operators' workload.

The frequent use of equipment led to increased fatigue of the hoses due to repeated bending and kinking. This created points of highly concentrated stress and fatigue in a specific part of the hose, which led to the delamination of its layers and localised material failure. An incident due to such delamination involved the failure and bursting of one hose during a fuelling operation. Fortunately, the failure was contained, and no other damage occurred, thanks to the safety measures in place, precautions and instructions that must be understood and applied during operation and maintenance to ensure personnel safety and protection of the equipment, and personnel training.

The incident investigation determined that all the mandated procedures had been followed, and the staff performed all operations as described in the operating manuals and checklists. Also, the safety valves and devices, both on the aircraft and the truck, worked as intended, and all systems had been properly maintained. However, the hose failed, and the question of why it happened so lingered. After studying the design specifications of the fuel trucks and the hose itself, it was determined that the one in use was semi-rigid and hard to bend (USDD, 2011). Its type conformed to the general specifications and standards; it had been widely used in fuelling trucks without any problems reported. Nonetheless, the investigation revealed a series of factors in this type of truck that contributed to the hose failure.

The finding pointed to a failure in the initial design of the hardware by the original equipment manufacturer (OEM) and a failure in the design of the operating and maintenance procedures. More specifically:

- This type of hose had a minimum bending radius requirement that made its reel large and heavy.
- The allotted space for the reel in the truck was barely adequate and did not make hose storage easy.
- Manual operation of the reel and the frequency of daily use were tiresome for the staff.
- There was no arrangement to make the hose's stowing position unique.
- There was no consideration about the weight of the hose to make it more easily handled.
- The checklist did not describe the correct way to stow the hose.
- There was no notification for the staff to avoid kinking the hose and fully reel it in to ensure it did not bend in the wrong way.
- The maintenance instructions required regular inspections with on-condition repairs, but no maintenance records were kept.

Indeed, during equipment design, the OEM followed standard practices and used generic specifications from the technical perspective. However, the OEM had not considered human engineering aspects (USDD, 2020) because the design did not consider the frequency of use along with the task difficulty and demands. It did not account for hindrances to operations due to the poorly designed stowage area and the reel size and weight. Moreover, the lack of a unique stowage position for the hose pressure coupling forced operators to simply deposit it in the stowage area, kinking the hose in the process. The manual operation of the reel was assumed to be manageable by the designer, without paying attention to staff workload and fatigue. There was no support scaffold to take on the hose weight during reeling. The combination of weight, the difficulty of handling, and the narrow space did not afford the staff time and physical and mental resources to attend implications from the reel operation or the hose stowage.

On the side of organisational procedures, staff was not trained to take special care of hose stowage. The inspection and maintenance schedule conformed to generic requirements but did not anticipate the increased wear and tear due to the high frequency and volume of fuelling operations. No inspection and maintenance data were available, meaning that no trend analysis of past failures could be performed. Technical risk management was minimal as it did not encapsulate additional or exceptional failure modes applicable to the organisational context of operations.

To give everyone their due, considering these failures, the organisation took steps to rectify the problem, as much as possible. Some of these steps were taken under my authority, including staff training and process improvement. Equipment modification and quality policies required getting the quality assurance and engineering departments involved. The result was to mitigate certain risks, avoid some, and accept those that could not be done away with. Not all hazards were economically feasible to address at the equipment level and, consequently, were addressed at the process level. Some risks were simply accepted as a cost of operations. More specifically:

- Equipment was modified to force a unique hose stowing position (poka-yoke[2]) (Evans & Lindsay, 2008), which prevented kinking of the hose.
- A scaffold was added to support the hose's weight while reeled and stowed, reduce the staff workload, and keep the bending radius within limits.
- Staff was trained to avoid bending and kinking the hose, making sure that they understood the hazards that these actions introduced.
- Notes were added to the checklists and documentation to outline the hazards arising during operations.
- The inspection and maintenance schedule included record-keeping for trend analysis.

The main lesson from this case is that when OEMs fail to include human engineering in their initial equipment design, problems will emerge during its operation. The equipment in use in aviation is costly, and its initial design and procurement might have taken place years if not decades before its current use. Not all usage cases

---

[2] https://asq.org/quality-resources/mistake-proofing

might have been considered, or the design scope might have changed, but not the specifications themselves. Regarding the fuelling truck, instead of making the system user-friendly, the OEM adhered to its technical specifications but made it cumbersome and tiring to work with. This led to staff not being able or willing to put extra effort to prevent hazards, which ultimately resulted in the failure of the hose. Consequently, the safety of staff and equipment and the health of the workforce were compromised by the initial design flaws, even though the organisation has had rigorous health and safety policies.

The second lesson regards the options when design flaws are identified. Although it makes a case for replacing them, this is not always operationally feasible or financially affordable. In these cases, risk management and process improvement tools and techniques can be applied to tackle the inherent hazards. Creating proper procedures and training the operators could be the most obvious risk controls, but engineering measures should precede. Each risk management strategy [i.e., avoidance, mitigation, transfer, or acceptance of risk (ISO, 2018)] has its own set of pros and cons relative to the hazards faced. Choosing the most effective one or combining any of the risk management strategies to improve operations can often be the determinative factor between failure and success.

CONCLUDING REMARKS

I believe the lessons shared through these two cases are applicable to any industry sector, regardless of the level of technology used and the type and size of operations. Ownership of tasks by the workforce should be one of the primary goals of management. Training employees to demonstrate a genuine interest in solving workplace problems early and removing hazards when first identified is an investment, with significant returns.[3] For this to happen, organisations must cultivate a work environment and culture where employees do not hesitate to report problems and know that the solutions that they suggest are given objective and unbiased consideration. Design of equipment and systems of work should reflect actual uses, and edge cases should be considered, as far as possible, to address hazards and manage risks. The end goal should be to learn how to succeed repeatedly and turn any failure into a success.

As Henry Ford said, "Businesses that grow by development and improvement do not die" (Ford News, 1923).

REFERENCES

EDA. (2021). *EMAR 21–Certification of Military Aircraft and Related Products, Parts and Appliances, and Design and Production Organisations (Edition 2.0). Military Airworthiness Authorities Forum*, European Defence Agency. https://eda.europa.eu/docs/default-source/documents/emar-21-edition-2-0-(approved)-30-march-2021.pdf

EU. (2014). *Commission Regulation No 1321/2014 of 26 November 2014 on the Continuing Airworthiness of Aircraft and Aeronautical Products, Parts and Appliances, and on the Approval of Organisations and Personnel Involved in These Tasks, OJ L 362, 17.12.2014, p. 1–194*. https://eur-lex.europa.eu/legal-content/EN/TXT/?uri=celex%3A32014R1321

---

[3] https://global.toyota/en/company/vision-and-philosophy/production-system/

EU. (2018). *Regulation 2018/1139 of the European Parliament and of the Council of 4 July 2018 on Common Rules in the Field of Civil Aviation and Establishing a European Union Aviation Safety Agency, and Amending Regulations (EC) No 2111/2005, (EC) No 1008/2008, (EU) No 996/2010, (EU) No 376/2014 and Directives 2014/30/EU and 2014/53/EU of the European Parliament and of the Council, and Repealing Regulations (EC) No 552/2004 and (EC) No 216/2008 of the European Parliament and of the Council and Council Regulation (EEC) No 3922/91 (Text with EEA relevance.). OJ L 212, 22.8.2018, p. 1–122.* https://eur-lex.europa.eu/legal-content/EN/TXT/?uri=celex%3A32018R1139

Evans, J. R., and Lindsay, W. M. (2008). *Managing for Quality and Performance Excellence* (7th ed.). Thomson South-Western.

FAA. (1974). *Aircraft Ground Handling and Servicing - –Advisory Circular (AC) 00-34.* United States Department of Transportation, Federal Aviation Administration. https://www.faa.gov/documentLibrary/media/Advisory_Circular/AC_00-34A.pdf

FAA. (2018). *Aviation Maintenance Technician Handbook–General (FAA-H-8083-30A). United States Department of Transportation, Federal Aviation Administration.* https://www.faa.gov/regulations_policies/handbooks_manuals/aviation/media/amt_general_handbook.pdf

FAA. (2021). *Aeronautics and Space 14 CFR § 23.2235-2245. Government Printing Office, United States Department of Transportation, Federal Aviation Administration.* https://www.ecfr.gov/current/title-14/part-23/subpart-C/subject-group-ECFR5f1d2bf5cc2f06c

Henry Ford (1923, Feb 15). Ford News, p.2.

ISO. (2018). *Risk Management – Guidelines (ISO/DIS 31000:2018)*, International Organization for Standardization.

Juran, M., and Godfrey, A. (1998). *Juran's Quality Handbook* (5th ed.). McGraw-Hill Companies, Inc.

NYC. (2014). *Fire code, Chapter 11.* New York City Fire Department, United States. https://www1.nyc.gov/assets/fdny/pdfviewer/viewer.html?file=Chapter-11.pdf&section=firecode_2014

PMI. (2021). *A Guide to the Project Management Body of Knowledge (PMBOK Guide)* (7th ed.). Project Management Institute. https://www.pmi.org/pmbok-guide-standards/foundational/PMBOK

USAF. (2008). *Technical Manual Ground Servicing of Aircraft and Static Grounding/Bonding (to 00-25-172).* Office of the Secretary of the Air Force, United States Department of Defense, United States Air Force.

USDD. (2011). *Detail Specification: Hose Assemblies, Rubber, Fuel and Nonpotable Water, with Reattachable Couplings, Low Temperature, General Specification for (MIL-DTL-6615G (W/ AMENDMENT 1)).* United States Department of Defense.

USDD. (2020). *Department of Defense Design Criteria Standard: Human Engineering (MIL-STD-1472H)*, United States Department of Defense.

USDS. (2021). *Foreign Relations 22 CFR § 120.38. Government Printing Office, United States Department of State.* https://www.ecfr.gov/current/title-22/chapter-I/subchapter-M/part-120/section-120.38

14 Stuck in a Holding Pattern

Human Factors Training Development for Sports and Recreational Aviation

Claire Greaves and Reuben Delamore
Tactix Group

CONTENTS

Sports aviation as a sector of the Australian aviation industry that covers almost half of the aircraft operations in Australia (Civil Aviation Safety Authority (CASA), 2021). As of 2016, sports aviation involved about 40,000 participants, more than 9,000 aircraft and 360,000 parachute jumps, and included a range of activities involving manufacturers, training facilities, competition participants, and enthusiasts. Sports aviation is varied and covers ultralight and weight-shift microlight aircraft, gliding, gyroplanes, hang gliders and paragliders, recreational unmanned aircraft (including models and drones), parachuting, warbirds, amateur-built and experimental aircraft, and recreational ballooning. Perhaps the most distinguishing feature of this sector is the different motivations to participate in its activities. Although many regular public transport (RPT) pilots, cabin crew, and other associated staff probably share the same affinity for aviation, aspects such as enjoyment, fun, performance, risk, 'rush' (Buckley, 2012; Pauley et al., 2008), or grabbing opportunities may not be the primary purpose for their engagement. These concepts associated with motivation are what sets sports and recreational aviation apart and are the crux of the challenge in creating tools for training and interventions for this sector.

Sports aviation clubs recognised that between 70% and 80% of their accidents involved some type of human error. Prompted in part by findings from a series of incidents investigated by the Australian Transport Safety Bureau (ATSB) as well as own investigations of the clubs, several sports aviation clubs began to realise their members needed training, and this training needed to address their unique risks. Although systems thinking would encourage operators to look more broadly throughout the organisation to resolve these risks (e.g., design and integration of equipment, systems, procedures, processes, and people), training and our human capabilities remain often the last line of defence that operators fall back on. This, however, assumes that training is effective in building knowledge and skills for all groups involved. To achieve such effectiveness, we collaborated with various sports aviation organisations to develop a series of e-learning modules. At that point, we were only tackling the knowledge aspect to help operators better understand their capabilities and limitations and what they can do to perform safely while enjoying their activities.

Interestingly, we noticed that the human factors (HF) and non-technical skills (NTS) training packages landing on our desks from various operators appeared to be similar, despite the wide and varied nature of the activities, motivations, and environments. This suggested that the nature of the risks that the operators were attempting to manage were different, yet the training was not. Our senior management had concerns regarding the high accident rate in the sports sector, and HF was perceived to be a significant gap contributing to the accident rate. For us and some of our supervising teams, there was a general feeling that HF training had become a 'one size fits all', where the application of key concepts (e.g., fatigue, stress, decision-making, and managing threats and errors) was not addressing the specific or unique activities that a particular operator or organisation performed. Indeed, the identification and integration of operational, individual, and environmental risks in the training programme seemed to be lacking. Hence, we asked ourselves *'Shouldn't the training be adapted to meet the unique needs and risks associated with each?'*.

To set the scene, the onset of NTS training within the aviation industry was primarily driven by a series of accidents and incidents that led to changes to international airline regulations. These changes arose from the recognition that pilots required training to combat key areas underpinning the series of unfortunate events leading to accidents (e.g., cockpit power gradients) and not just technical skills and knowledge. Training in NTS (also known as soft skills) in the form of crew resource management (CRM) has since been an integral and legislatively required[1] part of cockpit operations, even extended to cabin crew.

Most often, NTS training in the aviation industry focusses on those areas most critical to the safe operation of the flight, including crew coordination and cooperation, effective communication, decision-making, conflict and error management, stress and workload management, maintaining attention and vigilance, managing automation, and optimal situational awareness. However, although these skills are often ubiquitous across aviation sectors, their application is often not tailored to each context. Indeed, a focus on compliance with regulations might have led to poor application of training due to not considering specific risks of organisations or operational contexts and the required skills. There are a variety of activities and tasks performed, differing motivations or performance-shaping factors due to the operating environment, different-sized organisations in terms of personnel, and financial capacities to invest in training.

---

[1] See for example Part 119 of the Civil Aviation Safety Regulations 1998.

However, from a regulatory perspective, the level of guidance to develop training is limited by the sheer vastness of operations and unfeasibility of creating and tailoring advisory materials to everyone. As a result, CASA has invested its expertise in creating resources and materials that operators could use as a benchmark from which to tailor their training packages to their operational risks and needs (CASA, 2020). At least, that was the intention. What seemed to arise in industry training, however, was a uniform and homogenised application of key NTS concepts.

OUR OVERALL APPROACH

To begin, we needed to revisit the starting point to create or re-build an existing draft of a sports aviation training package. We worked with the internal communications and training development teams to understand the task and review what had been developed and presented previously. At this stage, the internal development team had been working in isolation from end-users, domain experts, and HF specialists, churning through the guidance material for HF training and reflecting that information in a presentation-style format. While we were sitting behind our office desks contemplating the development of this HF awareness training package, we realised that something was different about this group and their needs.

Our review of the initial draft developed by the training team identified what we were concerned about, another carbon copy program based on an RPT training model. We knew if we did not intervene and promote the tailored risk-based training approach (i.e., making the training applicable to their risks), it would not be fit for the users. RPT training models are based on operations of transportation that provide a service from place A to place B. They do not account for fun and games at point X in between those destinations that the audience of our sports enthusiasts were seeking. It needed to be punchy and hit the mark on the risks that the sector had identified due to recent incidents and near-miss events, as well as reflect what they were doing. Also, the training should be cutting-edge and engaging. We felt the traditional training development approach of leveraging from RPT organisations would neither capture what the key issues were for the domain nor the activities they were performing (e.g., not being multi-crew operations).

To bridge this gap, we requested the involvement of industry representatives in developing and reviewing the package through forums. We received positive responses from several sports and recreation federations and groups. Simultaneously, we reviewed the incident data of the ATSB, which was also bolstered with information from various sports and recreation federations. We found that published information was, at times, limited regarding the factors that contributed to the incidents. The review of the data and incident reports indicated that a generic approach to developing the NTS training, based on the guidance material alone, would not easily translate to sports aviation activities, environments, or equipment. The context of the operations was different. These aviators were enthusiasts who often built their own equipment and maintained it. Multi-crews were a rarity, and the knowledge base and performance capabilities of the aircraft and operators were wide ranging.

Therefore, the training needed to reflect the gaps in the intended user's knowledge and, most importantly, the nature of their specific risks. From our review of the data and information collected, five key areas were identified:

- Fitness to fly
- Situational awareness in aircraft operation and maintenance
- Individual and stakeholder communications
- Decision-making
- Managing the potential for errors and accidents

In the next sections, we describe two cases where we applied a risk-based approach to developing a bespoke e-learning training package for the sports and recreational aviation sector. We were enthusiastic to see how an innovative approach and slightly rebellious style of getting messages across could refresh training programs in this space.

ARE YOU FIT TO FLY?

One topic we wanted to focus on for the training package was the key areas of health and well-being, which were critical to impacting or derailing sports aviation enthusiasts. In our review of the literature and incidents sourced from ATSB, various sports aviation clubs, and CASA, we identified that health events (e.g., heart attacks) and health status (e.g., underlying chronic disease, such as diabetes) seemed to underpin some incidents/accidents. Within the broader context of sports aviation, the physical health requirements remain less rigorous than the commercial end-of-town possibly because the nature of the sports aviation sector and its risks are not uniform. For instance, older aviators, who may be more susceptible to age-related health problems (e.g., heart disease or diabetes), can be found more in some specific sub-categories of sports aviation, such as Warbirds.[2] Also, while medically prescribed drugs can play a role in performance, the use of non-prescription drugs can also impact performance.

Several clubs provided information and images from their incident databases with case study examples as a starting point to identify the need for a tailored training package. We wanted to provide a level of knowledge of how these factors could affect flight safety, and we focussed on the performance margins in this space. We reviewed various ATSB incidents and case examples provided by operators, and we extracted contributing factors associated with health for inclusion as topics within the module. Additionally, we reviewed the respective literature (e.g., age, fatigue, and drugs).

Moreover, having recognised earlier it was not about a 'one-size-fits-all' solution, while developing the training package, we had ongoing and regular consultations with the members from various groups (e.g., ballooning, gliding, warbirds, and parachuting) and respective regulatory inspectors. We also sought advice and insights from internal organisational medical specialists about demographics, the nature of issues witnessed, and their concerns. These parties provided review and feedback to

---

[2] https://australianwarbirds.com.au/

support the drafted modules and online versions before the latter was incorporated into the overall HF package for release.

Admittedly, the broad spectrum of sports aviation sub-sectors made it difficult to tailor content to each sub-category and make it exclusively relevant for the corresponding pilot profiles. We struggled to find a pertinent 'middle ground' to ensure user interest in or engagement with the module. Warbirds, for instance, are older aircraft requiring focussed maintenance, and their pilots typically resemble RPT operators with high knowledge and awareness of NTS and airmanship. These pilots are likely to be older but very aware of the effects of age and prescription medications on performance. Non-prescription drugs may not be as relevant to this group, just as age-related chronic health conditions may not be the most representative of the demographic of wingsuit operators.[3]

Perception also plays a pivotal role. Who is to say that the warbird operators are not all into non-prescription drugs? What about issues of dehydration or medical devices (e.g., pacemakers) and operating for long periods of time? Stereotyping and bias lead to generalisation about the persona of the operators. Those types of heuristics help in creating a general training course, but you do need to be careful. Self-report biases play a role in the perception of the type of people and the type of activity that they are undertaking and can lead us to assume and extrapolate about what they do or do not do in other aspects of their lives.

Considering these variables, we took the approach of a broad brushstroke with widespread sector-relevant examples: what is applicable or relevant to the wider group? That way, we were not removing content that could be useful and important to some but not others. We felt it would be best to be inclusive and discuss topics that were applicable to the wider audience (because we really did not know what people were doing in their personal time). Ideally, at the start of the training course, providers would be able to select the sports aviation sub-sector(s) most applicable to each trainee. Then, the modules presented would be those identified through the literature and incident analysis pertaining to the risk profile and operational considerations of that sub-sector(s).

Indeed, a refined approach would likely lead to better engagement and hopefully avoid people skipping through content in which they were not interested or felt does not apply to their specific training needs. This can be a key limitation of online training, hitting the 'next' button hoping to get to something useful or applicable. Unfortunately, at the time, we did not have the scope and time to tailor the material at that level, which would have enhanced the module substantially. Another principal constraint for the e-package was the method of learning (i.e., remote and self-directed) and genuinely engaging the audience without relying on facilitation skills in a face-to-face delivery mode. It was challenging to envision how we would do that. The principal challenge with e-learning is that you cannot present the material like a slideshow because there is no one there to provide the context. It must stand alone and do that well.

Moreover, although we engaged sufficiently with our audience and stakeholders, a more targeted integration plan for engaging users and having their input

---

[3] https://www.raa.asn.au/

weaved throughout the build phase would have facilitated a more robust solution. User engagement was attended at the drafting stage, but by then we, as creators, had already taken the content in a certain direction; perhaps, this rendered it more difficult for people to feel they could provide honest feedback. Next time, we would bring a variety of sector representatives into a physical or virtual room to discuss the big-ticket items and co-design the training modules in a more integrated manner. Also, from a user representation perspective, it would be valuable to include a wider cross-section of the group including various sector activities, ages, and experience levels. The more diverse the representative sample is, the more likely the product will meet their needs.

It is hard to answer the question about what worked well or not with this module. Nonetheless, we received some excellent feedback on the module content and presentation, and it was certainly well received by the club members. We do recall discussions around age and drugs. One could conceivably include a myriad of topics in an HF and NTS course. What we wanted to target was this idea of a 'risk-based approach', which was not seen in this aviation sector at the time. Rather than grabbing generic topics of HF off a website or toolkit, our approach required drawing upon general information and working with users and stakeholders to determine what is relevant, of high priority, or perceived as presenting the highest risk to operations, the safety of others, and performance. From a consultant perspective several years ago, it would be interesting to know whether the sector is still using this training as part of their club memberships and if not, the reasons why to inform and enhance future training development.

Reflecting on this experience, there are several key takeaways. What tops the list is to be mindful of perception and stereotype bias. Speak to your audience, the users, stakeholders, and subject matter experts (SMEs). Instead of making assumptions about the group, engage them early and throughout the project to ensure that what was agreed upfront made its way into the final design/deliverable. Secondly, capture user needs and issues using a register. This allows us to trace the original problem and how we have addressed it by way of our intervention. Finally, touch on important topics and use data to inform the focus areas. It is essential not to get caught in a 'user-led' approach. Minimise influences by others' views on what the sector or group is doing and what it faces. Stakeholder engagement is key, and different stakeholders, including users and SMEs, can have divergent perspectives. User-centred consultation is critical, and understanding the activities and tasks they are performing is fundamental as a skill set and tool to HF practitioners.

MANAGING THREATS AND ERRORS, RUSH, AND RISK

To help navigate through this part of our chapter, we first need to delve a little into the background of two key models: Threat and Error Management (TEM) model and adaptations on the Dynamic Safety Model. As a concept and tool, TEM originated from the Line Oriented Safety Audit (LOSA) program, a collaborative partnership between Delta Airlines and the University of Texas.[4] LOSA was originally developed

---

[4] https://www.faa.gov/about/initiatives/maintenance_hf/losa/history

to evaluate the performance of NTS training, through the observed application of CRM skills and behaviours among RPT flight crews. The audit tool was later developed to capture crew errors and error management, and the TEM framework evolved further, benefitting from research and the developing understanding of the interplay between human performance and safety within operational environments. Key contributors to this broader research included Rasmussen (1980, 1982, 1997), Helmreich (2000), Helmreich and Musson (2000), and Reason (1990, 1997). Their work generated a deeper understanding of error, risk, systems design, latent failures and human performance at individual and team levels.

The TEM model is usually represented as a split triangular shape, comprising at least three different levels of process and strategies to manage or mitigate threats and error. Its structure includes two axes with time on the vertical axis (y) and available recovery options on the horizontal axis (x). Standing on its point, the TEM model represents a process that flows from the top to the bottom. The presence of threats and risk above the model are deflected away by threat and risk management strategies. The inception of a threat passing through these strategies signifies a failure. If the threat continues through the preventative mechanisms, it may manifest into either a systems failure or a human error. At the error management layer, these errors are 'trapped', 'mitigated', or 'avoided' relying on action and intervention from the crew or other supporting systems, such as automation, traffic collision avoidance systems, and other corrective measures. The TEM framework and process consider an operation in four stages:

- Pre-planning,
- Day of operations management,
- Prevention of undesired aircraft states, and
- Consequence management or mitigation.

These stages are applied across all flight phases (e.g., pre-departure or pre-planning, departure, climb, cruise, descent, and landing). As such, TEM utility is through the preventative management of threats, risks, and errors by using a standardised systematic approach with predetermined procedures to mitigate and manage adverse consequences. If the preventative mechanisms fail to manage the event, the operational team members are then the 'final filter' and are required to manage the situation to mitigate the consequence of the event. To support this process, CRM skills are seen as integral tools to support human performance against the potentially adverse impact of threats and errors.

In the model, time and options are finite and reduce in availability and utility the further the crew progresses through the situation. This reflects the real environment; for instance, commodities and protections, such as fuel, or the altitude that the aircraft is clear above local terrain and structures, become gradually limited with time. The earlier in the process any issues are captured or managed, the more time and options are available for operators to utilise to manage an unfolding situation. Critically, TEM as a process presumes a sequential handling of threats and errors with the benefit of pre-considered defences and management strategies. Realising this was key in developing our TEM section of the training package; the history and development of TEM indicated that its traditional form could not match sports aviation well.

The problem was that TEM had been a cornerstone training subject for commercial aviation since the 1990s, and our management had assumed this would also hold true for sports aviation. For our NTS/CRM package, we needed to ensure the intentions of TEM were supported with practical steps, strategies, and tools that aligned with the operators' needs and capabilities, their equipment, their risks (if known), and their operating environment. This training module had been left last for development due to the complexity we had seen when reviewing accidents and incidents. However, we were also struggling to visualise how traditional TEM could be applied to the unique risks of the sports aviation domain and articulated in a meaningful way to the audience. It had become clear that incidents were not often due to the type of systemic and latent failures typically found in the more technologically advanced team-based operations of commercial aviation. The sports domain included enthusiasts who often had a hand in building their own equipment and maintaining it. Multi-crews were a rarity, and the knowledge base and performance capabilities of aircraft and operators were wide ranging. This differed to what was seen in RPT organisations where the NTS training reflected examples focussing on shared failures within multiple crew member interactions, advanced aircraft technology, structured maintenance-based activities, and other detailed investigations into contributory factors.

By comparison, training and education for sports aviation appeared more cursory. It was common to see sports aviation using the same RPT incident examples in their training. Yet, based on feedback from domain stakeholders, these examples were not as relatable to the type of operations, actions, or events of sports aviation. Incidents in this sector were often events involving a single person (e.g., single pilot vs the multiple crew members and maintenance support seen in RPT) or based on smaller teams with less formal training and procedures. Weather and terrain events (e.g., flight into terrain or losing visual reference in cloud and losing control) were frequently identified as contributory factors linked to the operators' decision-making to continue to fly in bad weather or not follow a previously planned route. These issues were minimised in RPT operations due to the height above terrain and the supporting technology. Many of the risks that would effectively be unacceptable in RPT were often not even considered a risk by some sport operators (e.g., wingsuit pilots and their proximity to terrain). Also, the more reactive nature of the operations (e.g., flying at a low level in an area previously not known, to fulfil an opportunity) can present risks the pilots are unaware of (e.g., powerlines or terrain may lie ahead), which, interestingly, can even contribute to the gratification of performing the activity.

The TEM concept, of course, has relevance to flying and operations in the sports domain, but in its traditional and accepted form, it was not so clear cut for areas like glider competitions, aerial displays, wingsuit activities, parachutes, balloons, and hang gliders. Even a cursory review of the history and development of TEM reinforced that the expected behaviours and motivations of RPT crews may not correspond with the nature of sports activities (e.g., opportunity fulfilment, competitive behaviour, split-second timings, un-tested changes to plans, increased exposure to the risks of terrain, weather, and other unknown factors). In short, not only were we dealing with environments, technology, and levels of training different from RPT but also some inherent sports aviation elements seen by RPT crews as the 'wrong stuff' (Moore, 1997). While some areas had similarities, we were dealing with a potential misfit of TEM to sports aviation due to the nature of the operating environment and the differing goals of the sports aviators from RPT operations. So, where did TEM fit in? After collating the information, we presented our findings to representatives of the domain and asked them that same question. The resounding response from our industry partners and end-users was that TEM was not a good fit in its standard form.

It was time to go back to basic principles, which is something we should do more often as professionals to make sure we do not blindly apply guidance out of the intended context. This activity would help us identify the limitations of the model for the sports domain and define the problem. We wanted to find a way to tailor TEM to the tasks and needs of sports aviation users. Our preliminary findings of the accident and incident data and literature review indicated we should further investigate three phenomena: firstly, the propensity and individual appetites for risk aversion and tolerance (Pauley et al., 2008); second, the excitement sometimes linked to rush at the limits of individual capabilities (Buckley, 2012); and thirdly, competition. These drivers and motivators appeared to be key differences between sports aviation and RPT.

We needed to find a way to incorporate these phenomena into the traditional TEM concept. Earlier discussions with end-users and the information from our literature review had also prompted the recollection of a presentation provided by the Royal Aeronautical Society and the London Metropolitan University in 2005 based on Rasmussen's (1980) work on systemic accident analysis and Cook and Rasmussen's (2005) work on safety and drift. The presentation had illustrated accident causation as dynamic and constantly in flux, an approach that we needed to explore as a possible alternative or enhancement for the TEM package.

Although the Dynamic Safety Model (Cook & Rasmussen, 2005) primarily considers accident causality, it can also be used to find out why accidents do not happen and then foster the repetition of those activities. This line of thought is also aligned to more modern theories than older causation models that contributed to the development of TEM. The Dynamic Safety Model incorporates the concepts of risk, human performance, and external influences that may contribute to a drift of the operating point towards failure. Figure 14.1 presents our attempt to simplify the representation and complexity of influences within the operating envelope and in relation to the boundaries of the original Dynamic Safety Model.

The operator/operation is represented as a point/area inside an envelope of safe operations and a wider 'bubble' of acceptable performance. This point is dynamic, moving within the operational envelope due to forces and counter-forces directed on it from factors related to the safety, efficiency, human, operational, and other performance boundaries of the bubble (e.g., social expectations). At the same time, such forces are created within the operational area itself (i.e., because of the individuals making up the team and aircraft performance). Therefore, the entire operational envelope is shaped by several layers that represent thresholds for failure and act as sources of 'pressure' (e.g., legislation, standards, policies, technical, and weather

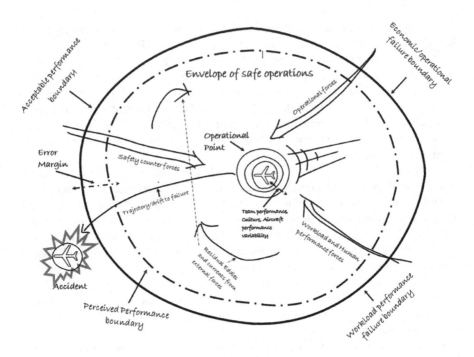

FIGURE 14.1 A simplified representation of Rasmussen's (1997) Dynamic Safety Model reflecting both a safe operating point and an operating point that has drifted to failure, breaching the boundary of acceptable performance.

limitations) that act on the system to deter it from reaching and crossing the acceptable performance boundaries.

The boundary of acceptable performance provides the threshold beyond which errors can visibly and detrimentally affect performance. At this boundary, environmental, organisational, and societal opportunities and constraints create currents that interact with local operational factors and increase or decrease the distance from the boundaries and affect the position of the operational point. Examples of 'safety currents' can be safety programs, training initiatives, or respective technology (e.g., equipment capabilities, information provided, alerts, and warnings). Conversely, routine violations and workarounds (i.e., habitual actions that breach rules or proposed norms but are accepted as normal work and usually related to poor design or poor procedures) may contribute to a reduction of the operations and boundary distance.

The economic or operating boundaries represent elements such as the motivation to cut costs and the duration and quality of tasks. The human performance boundary relates to the task demands and human capacity. If the system pushes for optimisation, the effort of the operators maximises to get more time for less. Failure at this threshold can occur due to overload, leading to mental and physical fatigue, and underload or workload, creating conditions for complacency or limited vigilance. The premise is that while the operating point moves in response to all forces and currents, if it remains within the bubble, the operation is not likely to result in an accident. The higher the distance from the boundaries, the lower the risks of failure (Morrison & Wears, 2022).

Hence, acceptable performance can represent the combination of various interconnected aspects, such as personal performance, goals or motivations, safety, or even equipment performance. The margin between the envelope of safe operations and acceptable performance boundaries represents error tolerance, which can be engineered (e.g., safety margins designed in equipment) or perceived. Conceptually, this margin represents the last opportunity of operations to default to their safe operational envelope before they reach unacceptable performance (e.g., accidents, economic crashes, or serious health conditions).

However, the perceived tolerance is not static; it can be reduced or increased depending on specific events. For instance, when operators breach the margin and there is no negative consequence, the margin could be reset at a new point closer to the acceptable boundary and, thus, reduced. Conversely, an accident experience might lead to an increase in the error margin, meaning a narrower 'safe envelope' and efforts to move the operational points towards its 'safe centre'. Simply put, the closer the operating point to the boundary of the envelope, the less the available defences and tolerance. If the failure boundary is crossed at any point, the system has drifted into a state with adverse outcomes.

Cook and Rasmussen (2005) later developed the idea of coupled systems within the same model. The coupling uses controls for risk and performance that can manage the influences from the boundaries to a specific extent, shifting the operating point from a high- to low-risk positions within the performance envelope far from the boundary of acceptable performance. In RPT, maximum distance of the operational point from the thresholds of performance is achieved through counter-currents generated by training and standards, procedures, automation and aircraft performance, positive crew actions, risk and error management strategies, etc. The expected result is the safe, planned, mostly predictable, and efficient flight journey, which is primarily driven by a goal to minimise human interference using automation and management of flight parameters to minimise costs (i.e., fuel and maintenance).

The TEM model in RPT supports the ability of crews to achieve this goal through high-performing and trained crew operations and standards and embedded risk and error tolerance through operational procedures and the use of technology and equipment (e.g., automation, fuel management, and aircraft performance monitoring). Employing the TEM framework, however, requires significant time and resources to determine what risks may exist and then develop plans to mitigate those risks. Employing mechanisms like the ones mentioned above, even scaled down, in smaller, mostly single-operator environments, can present key financial or business challenges. While safety is far from an undesired state in sports aviation, it was often the presence of a challenge or competition, the demonstration of skill and elements of fun that attracted the operators to pursue their activities.

The utility of the Dynamic Safety Model representation helped to make sense of how RPT and sports aviation environments are fundamentally different. In sports aviation, the operating point can reflect an already high-risk system, away from its 'safe centre' (Severinghaus et al., 2012). While some groups such as Warbirds reflect a tightly coupled high-risk operation (e.g., through procedures and planning), the system tolerances to the margin of error and acceptable performance boundary are minimal; in part, this can be attributed to risk and rush and the motivation of putting on 'a good display'. Environmental factors alone push the operating point closer to failure through reduced clearance from terrain, more exposure to weather, and reduced performance capabilities of equipment and crew.

After we understood the differences between the two sectors, we started developing a TEM approach that focusses less on system-based preventions and more on active mitigations operators could consider. In pitching this concept as an alternative to the traditional TEM model, the end-users and stakeholders agreed that the divergence between their activities and RPT stemmed from the following features:

- The primary motivator for the activity is for fun, not commercial gains as in RPT operations.
- It is about experiences and opportunities, not necessarily getting from one point to another. Opportunities can arise at any time, and pilots have the freedom to change plans.
- In some environments, there is a thrill in pushing their aircraft limits, challenging themselves, and experiencing the adrenaline of flying.
- Competition flying may prompt the need to put into place a plan last minute without the opportunity for a detailed review to account for risks and unknown factors.
- Pilots are more affected by the environment they are flying in, both physiologically and in proximity to the terrain.
- In some environments, there is a goal to 'put on a show'. This requires significant planning and communication, while also maximising the audience's perception of risk and thrill.

Coupled with confirmation from our stakeholders, we felt that we had enough of an understanding of the differences among activities, their risks, and the applicability or not of the models for the different operational environments. While not always actively seeking risk, nor with an intent to be exposed to risks, many of the activities undertaken in sports and recreation aviation naturally had a reduced margin or tolerance for error. While safety was a desired outcome, that is not the same as saying it was the goal of the activity. Thus, we merged efforts to reframe the intent of the TEM model to consider these factors, focussing on awareness and the use of practical tools and strategies to manage or mitigate the risks related to the goals of sports flights (competition, enjoyment, etc.), risk perception, reduced separation from structures and environmental threats, and the reliability and capability of aircraft.

Upon reflection, we evaluated a unique environment in a structured way and involved end-user and industry stakeholders in the module design, who were now championing its value. We had achieved what we were trying to get other operators to do, that is, tailor training to meet and address their own tasks and activity-specific risks. But we made one mistake; we had not involved everyone. While working on this project, internal stakeholders had shifted and reorganised, and our reporting lines changed. We had not been as judicious as we should have been due to infrequent, and, in hindsight, ineffective communication with our internal stakeholders.

Our previous reporting lines had been directed into an area with a detailed understanding of what we were trying to do and familiarity with the concepts we were managing. While for us, the understanding and need to apply TEM in a tailored way were clear, we struck a challenge to explain the concepts we were addressing to team members not familiar with the area. There was distrust in the new approach as TEM was normalised by many as the way to manage any operation. Who were we to change such legacy training content? We began to feel under-supported by the internal team primarily because of the difficulty we had in explaining the customised model in a simplistic and accessible way (i.e., non-HF jargon).

We were under immediate time pressure to deliver the final product for internal and external stakeholders. The time limit was internally imposed and linked to a perception that the models and concepts were solely our academic musings, and there was a need to correct the approach back to TEM as applied to RPT. Consequently, the intended content of the module was lost as the focus became achieving delivery, re-writing the section in line with TEM legacy training, and managing internal stakeholder perceptions. At the time, the concepts made sense to us and were based on sound HF principles, research, and user engagement. However, the concepts did not make sense to everyone else. The management team had no appetite to review the suggested approach nor the research and information that had taken us there. Our failure to communicate effectively and manage internal stakeholder expectations let us down. Eventually, we made them understand that we had not invented the approach or made it up. Then, the feedback we received was simply that our approach was 'too new', 'too novel', and not widely accepted to be published by the organisation. The experience took some time to get over.

There are some great lessons learned here for all HF practitioners. We were too caught up in our focussed enthusiasm for what we were working on; at the time, we thought success was eminent. Had we applied a more proactive internal stakeholder management process and incorporated critical thinking about key organisational psychology principles around organisational and culture change, we may have better articulated our approach to the broader audience and enjoyed a successful outcome. In the early roadshows, where we had been communicating what we were developing for the sector, we had been using language broadly and deliberately as a tool to garner support from the sports aviation groups. We achieved buy-in by reinforcing the general view that the domain was different. Although the concept of TEM was not wrong for this sector, it simply fell short of reflecting the experience of the end-user group. We were aiming to enhance the model and take a novel approach to expand it.

Another key takeaway from the outcome of this module was to be aware of one's areas of expertise and knowledge and know when to park the ego. It is easy to be frustrated and disappointed when you or the team have been working tirelessly to deliver a quality piece of work and the feedback that you receive is less than stellar. It is a mistake to lean away from respectful interactions with broader stakeholders. One needs to lean in to understand another's perspective and position so that all parties are on the same page and work towards the common goal. Rather than 'throw out the baby with the bathwater', it may be as simple as re-articulating the message. For us, our message should have been '*The concept of TEM works, we just need to tailor it for this domain'*. Such a statement may have been effective at working through the miscommunication or misinterpretation that our message created, '*TEM is wrong, and we are going to invent something new to address it'*.

Although we did not achieve the delivery of the module that was desired by the industry stakeholder groups, and the final product deviated from our initial planning, the relationship building and buy-in with the sports sector was excellent. It was a key example of a strong partnership and collaboration towards achieving a training programme that would work for them even if this meant it would not incorporate our original approach. From our experience, these lessons have taught us how to approach the problem of applying novel theories to long-standing problems in a more integrated, research-focussed, and collaborative way. The overarching success of our work is that an e-learning training package was developed and provided to a unique sector of the aviation industry, and the application of existing theories and models can be adapted to novel contexts and not just 'the big players'. What stands in the way of any successful training or HF intervention is that organisations and practitioners alike need to truly understand their specific risks within their environment so that they can tailor how they address these areas. Without this capability, no training or intervention will suffice.

In summary, what we can advise HF and other professionals based on the learnings from this case spans the following areas:

- Perform a deep investigation through proper analysis and research into an area. This will help to understand what is important to the actual operators/ users as to what they think they know and identify what they need to know.
- Challenge the norm. Specialists and practitioners will encounter difficulties when breaking the mould or challenging the accepted thinking/models/ theories, especially within highly regulated environments.
- Avoid the glass tower. Language is key to successfully communicating a message, but it can also be a key area of discomfort if not in line with the norms.
- Apply it at your own risk. It is generally believed you can take any commonplace HF tools and apply them everywhere, in every situation. However, tools should be selected based on their suitability to the environment and account for the specific risks of that environment and operations.
- Deeply understand the content and underpinning science when communicating potential changes for legacy knowledge. If you can communicate a concept simply and with real-world examples, you will have more success than us.
- Involve everyone. One pitfall of being part of a specialist area is that the cross-pollination and collaboration among different teams or divisions within an organisation can be limited. It is important to engage widely, to consult with SMEs to ensure the suitability of the delivered product, and to leverage the knowledge of those in other industry areas. Remember your internal stakeholders, like, really remember them and bring them all on the journey with you.

REFERENCES

Buckley, R. (2012). Rush as a key motivation in skilled adventure tourism: resolving the risk recreation paradox. *Tourism Management, 33,* 961–970.

CASA. (2020). *Advisory circular, AC 119-12 v1.0, Human factors principles and non-technical skills training and assessment for air transport operations, Civil Aviation Safety Authority, Australia.*

CASA. (2021). *Sports aviation, Civil Aviation Safety Authority, Australia,* https://www.casa. gov.au/aircraft/sport-aviation/about-sport-aviation (accessed August 2021).

Cook, R., & Rasmussen, J. (2005). "Going solid": a model of system dynamics and consequences for patient safety. *BMJ Quality & Safety, 14*(2), 130–134.

Helmreich, R. L. (2000). On error management: lessons from aviation. *BMJ, 320*(7237), 781–785.

Helmreich, R. L., & Musson, D. M. (2000). Threat and error management model: components and examples. *British Medical Journal, 9,* 1–23.

Moore, J. (1997). *The Wrong Stuff: Flying on the edge of disaster.* Specialty Press.

Morrison, J. B., & Wears, R. L. (2022). Modeling Rasmussen's dynamic modeling problem: drift towards a boundary of safety. *Cognition, Technology &Work, 24,* 127–145.

Pauley, K., O'Hare, D., & Wiggins, M. (2008). Risk tolerance and pilot involvement in hazardous events and flight into adverse weather. *Journal of Safety Research, 39,* 403–411.

Rasmussen, J. (1980). Notes on human error analysis and prediction. In *Synthesis and analysis methods for safety and reliability studies* (pp. 357–389). Springer, Boston, MA.

Rasmussen, J. (1982). Human errors: taxonomy for describing human malfunction in industrial installations. *Journal of Occupational Accidents, 4,* 311–335.

Rasmussen, J. (1997). Risk management in a dynamic society: a modelling problem. *Safety Science, 27*(2–3), 183–213.

Reason, J. (1990). *Human error.* Cambridge University Press, New York, NY.

Reason, J. (1997). *Managing the risks of organizational accidents.* Ashgate, Hants, England.

Severinghaus, R., Cuper, T., & Combs, C. D. (2012) Modeling drift in the OR: a conceptual framework for research. *Simulation Series, 44,* 9–16.

15 Undertrained Workforce and Poor System Designs

Jose Sanchez-Alarcos Ballesteros
Factor Humano

CONTENTS

LACK OF TRAINING

This first case dates back to 2007 during my work at a nuclear power plant (NPP). The plant had an outstanding safety record that was boasted in industry as an organisation in which almost three years had passed since the last non-programmed reactor stop ('*scram*', in the nuclear lingo). However, NPPs were, and still are, the subject of political unrest. Despite their safety records, names like Chernobyl,[1] Harrisburg,[2] or Fukushima[3] were widely known, while, at the same time, some key features of these cases could be conveniently forgotten (e.g., intrinsically unstable designs or susceptibility to extreme natural phenomena).

Until 2004, in Spain, NPPs enjoyed support to navigate the concerns of political and ecological groups. However, newer governments became openly hostile to the nuclear energy industry. In the emerging political climate, opening a new plant was difficult, as was sustaining those already operating. Existing plants were under the threat of being shut down. The owners of the plants, conscious of this phenomenon, decreased their investment in the future of NPPs.

In some cases, political fighting reached an emotive pinnacle, culminating in the pronouncement of a 'non-nuclear city', the place where more than half of the NPP workforce was living, although the NPP was the main economic engine of the region. NPPs have been targets of demonstrations by ecological groups, including assaults to NPP installations.[4] In 2004, a well-known journalist made a TV programme about the problems with NPPs, with references to Chernobyl.[5] The journalist attributed cancer to an NPP because it appeared in three consecutive generations of a family

---

[1] https://en.wikipedia.org/wiki/Chernobyl_disaster
[2] https://en.wikipedia.org/wiki/Three_Mile_Island_accident
[3] https://en.wikipedia.org/wiki/Fukushima_nuclear_disaster
[4] https://file.ejatlas.org/img/Conflict/2216/garona-cierre-ya-4_1_.jpg
[5] https://vertele.eldiario.es/videos/actualidad/mercedes-mila-diario-centrales-nucleares_1_7787551.html

DOI: 10.1201/9781003349976-15

and they lived close to a plant, although there was no mention of the possible contributing genetic factors.

In September 2006, a mainstream newspaper published a front-page headline about a plant that was about to close because its operating licence was denied an extension beyond 2009.[6] The same source mentioned government intentions not to consider an extension of the 40-year useful life of NPPs, thus indicating a clear threat to the existence of the remaining plants. That was the political and media environment of that period, and it can explain some things that would happen later. Things changed after the public announcement about this plant closure, especially since it was presumed that the newspaper had shared credible information from government sources. After the alarming news release, the NPP suffered five scrams in a few months, while almost three years had passed since the last one. Something was clearly wrong at the plant, and they wanted to know what the issue was.

Due to the hostile socio-political environment, the management of the NPP was aware that their practices were closely examined. Among operational concerns, they were working on the development of transparent job descriptions to outline the key competencies of each role. Considering the unfolding situation and political and industrial unrest, the plant owners were decreasing operating costs, and hence, there was a moratorium on workforce growth. Subsequently, the NPP contemplated the recruitment of temporary workers and subcontractors, although, as explained below, their work description would include the performance of traditional, ordinary operational tasks.

In the past, when the reactor necessitated fuel recharge, the plant could cease its operations for four months, and the substantive workforce would perform all the required tasks, usually with considerable overtime expenditure. It was said that when the plant stopped, workers could buy a new car with the overtime payments. That solution was expensive for the NPP, mostly due to the extended downtime. Hence, major changes were introduced to reduce shutdowns to under one month. These changes led to the massive recruitment of temporary workers.

During downtime, temporary workers were subject to surveillance by members of the permanent workforce who were knowledgeable about the task and their related safety issues. Nevertheless, most of the safety incidents in the plant happened while it was closed, and they were attributed often to the inexperience of new, temporary workers. It was common for jokes to be made about inexperienced workers, like that they would get minor radiation traces that, once detected by the sensors, would prevent them from leaving the plant until they were decontaminated by members of the medical service dressed in hazmat gear – quite an alarming fashion.[7]

Despite these safety challenges, the practice of using temporary workers was incrementally extended from downtime periods to ordinary operations. Eventually, these workers became a common part of the employment landscape, not restricted to downtime. However, it was a major difference in the legal side: justifying a temporary contract was straightforward during downtime, and legal complications were

---

[6] https://elpais.com/diario/2006/09/17/sociedad/1158444001_850215.html
[7] https://img.joomcdn.net/a0239125a608a896a3019849899decad751573e4_original.jpeg

not expected. That situation would change when temporary workers were included in the normal operations phase.

When temporary workers became involved in routine operations, rather than just seasonal downtime, they acquired more experience and familiarity with operational routines. As such, surveillance strategies relaxed. Additionally, justifying why normal operations are performed by temporary workers is harder than that in the downtime situation. An increasing pressure to convert them into permanent employees should be expected.

That was the situation when the safety incidents started to happen. It was hard to explain why almost three years had passed without a single scram, and they now suffered from five in a short period. Although NPP managers could intuitively guess that the media headlines and the resulting anxiety across the workforce could be related to this, they wanted to know the technical side, that is, why did these scrams happen? I was invited to participate as an external expert in the effort to answer this question since I was well-known in the organisation. This was because of my previous work with the NPP in the organisation and human resources field.

The session was opened by the operations manager, who explained the technical facts to team members. In each case, he concluded with a 'lack of training' explanation as to why the scram happened. My limited knowledge about the technical aspects of an NPP did not allow me to challenge the explanations. However, the 'lack of training' label was surprising, especially when it was used repeatedly.

Inescapably, the explanation was seeking a question that, in turn, would require a clear answer. Why were the tasks leading to the *scram* performed by workers without the right training? All the attendants with a technical background were surprised by this question that could appear as rhetoric or a simple *boutade*. However, they caught the relevance of the point: the problem was beyond the mistake of a worker in every instance. Something was wrong at the organisational level.

In some way, I was playing safe by raising concerns about the 'lack of training' issue because of my previous experience. At that time, I was familiar with air accidents that were attributed to this cause and involved stakeholders who attempted to move the focus away from design or organisational pitfalls. The final operator could become the breaker of a system and the stoppage point to mask the need for major system changes (Ballesteros, 2007). Although system design plays a significant role in the identification of wider issues, in this case, lack of training was real and legitimate; people really were performing tasks without the qualification that would allow them to do it! Why this organisational pitfall?

At this point, I must add more context. Labour regulations in Spain were, and still are, extremely protective. There are several cases where a temporary worker, supported possibly by a trade union, can sue a company, forcing it to recruit the worker as a member of its permanent workforce. Severance pay is high, and the reasons for firing permanent workers must be carefully justified. Therefore, if the company is not able to justify why an employment is temporary, the worker can become a member of the permanent workforce through the decision of a judge. One of the arguments that could be used to advocate that a worker should become a permanent employee is the attendance of training sessions. Normally, training is provided to permanent workforce members. Therefore, the Labour Relations Department advises managers not to include temporary workers in training sessions, unless they can justify that training addresses a specific activity related to the objectives of the temporary contract.

Hence, once the 'lack of training' issue was raised, every piece fell into place. Managers were using temporary workers without differentiating them from the permanent workforce, except for aspects that they had been advised to avoid the legal risks. One of the industrial relations risks was training provision. Recruiting temporary workers was common practice, but providing training could strengthen the allegation that the employee should be made permanent, with ongoing benefits. As such, those workers could claim different work conditions. Apart from the safety briefings offered to anyone visiting the plant, additional training was not offered to temporary workers.

This situation made clear that something should change. Subsequently, tasks were analysed to identify safety-critical ones, meaning those that should be performed, or at least closely supervised, by members of the permanent workforce. Those tasks, if incorrectly performed, could contribute to physical or radiological damage. Therefore, they were subject to close supervision if not directly performed by experienced workers.

Notably, working inside an NPP, even on a temporary basis, has some interesting features. Safety is always a concern and intrinsic to operations, but it is also an objective, and the achievement of which must be convincingly demonstrated to external stakeholders due to political pressure. For instance, to avoid false accusations, dosimeters (a device to measure the uptake of external ionising radiation) remain inside the NPP; every worker must wear them inside and doff them before they leave the plant. In that way, it is possible to differentiate the radiation levels accumulated in the NPP from that which might come from exposure to other sources. The possible attribution of any sickness, mainly cancer, can be examined accordingly.

This 'safety-show', as I named it, necessitated time-consuming induction training for everyone who entered the installation on a permanent or temporary basis. Hence, those sessions could have been the right forum to introduce critical information with the aim to avoid major mistakes. Still, permanent workforce members were encouraged to keep overseeing and mentoring the temporary workers who, despite some of them being in the plant for a long time, had not received full training. In some way, the plant had suffered a drift toward a decreased safety level, and the five consecutive scrams made all the involved people more careful.

However, this initial effort was made by the experienced workers of the plant, without organisational guidance beyond a 'be careful' prompt. The identification of the organisational issue beyond the 'lack of training' should ideally drive the establishment of standard criteria about who can and cannot perform safety-critical tasks. In other words, the tasks linked to every specific job description could have been reviewed and separated into tasks that could be delegated and tasks that should be performed by the jobholder. In that way, the decision about what could be delegated would not anymore be left to the judgement of each worker, and the organisation would maintain this under control. Nonetheless, the plant kept working with the abstract 'be careful' mantra addressed to experienced workers and relied on their supervision practices. This mixed solution revealed itself to be enough to stop the sequence of scrams.

Indeed, the most reasonable solution should be to hire on a permanent basis those required to run the plant and provide them with all necessary training. However, once the political hostility was clear and public and the plant closure was more than a remote threat, recruitment became more tightened. The owners did not want to pay expensive severance payments if they could avoid them, especially as they knew that the plant was to be closed and the workforce would be fired at a high cost. Adding new members to the permanent workforce would mean an increase of a cost that, at that moment, was a real threat.

On the other hand, temporary workers were present in the normal operations well before the scrams. Why did the incidents not start then, and why did they appear only after getting notice of the closure of the plant? There is not a clear answer to that, and we can only launch a hypothesis, partially based on my impressions from my visits to the plant before and after the alarming increase of scrams. Before the publication of the newspaper article about the future of NPPs, permanent workforce members were relaxed about their own job safety. Compliance with the surveillance requirement was informally adopted and performed in a natural way, and despite the lack of formal training, experienced staff could explain to the temporary workers why the tasks were designed in a specific way and the risk of doing them in different ways. After the article was published, permanent workers possibly felt threatened, and their minds were not precisely on the necessities of the temporary workers that they were supposed to watch but in their own insecure situation. Therefore, the work environment could have become especially tense and prone to errors. Increasingly, more operational tasks were delegated to temporary workers lacking appropriate training or adequate work experience and not consistently supervised.

Within the limitations of a hostile external environment that, despite the sky-rocketing energy costs, is still thriving,[8] I consider the improvement explained above a successful case. Admittedly, the solution was limited since some major issues are untouchable, and it was retrospectively applied after the NPP experienced the scrams. Nonetheless, improvements were made in two ways:

1. Improving the induction training, that is, the knowledge that anyone working inside the NPP must receive, regardless of whether the collaboration is temporary or permanent. This training is centred on self-protection and basic knowledge about radiological risk.
2. Improving the task distribution by defining in every job description the non-delegable tasks where an error could drive a serious safety problem if performed by a temporary, less-skilled worker. It implied the inclusion of a new section in the job descriptions, making clear which tasks could be delegated and which ones could not.

Despite these options being enough to stop the safety incidents, they were far from being an adequate solution, which necessitated a focus on system design and deeper and wider organisational changes. However, this is what was feasible in the legal

---

[8] https://elpais.com/economia/2021-11-11/espana-se-desmarca-de-francia-y-sigue-con-el-plan-de-apagar-la-energia-nuclear-en-2035.html

environment of Spain. Certainly, having more flexible laws about the workforce, that is, hiring and firing according to workforce requirements as well as affordable severance payments, would make everything easier, including full training and being in the permanent workforce.

In the playground of politics, someone must acknowledge that barriers to firing also become barriers to hiring; that was the external trigger leading to the internal organisational issue. Therefore, the options were defined in very limiting terms. Inside them, safety issues were solved, despite the organisational conscience about the existence of a better solution that, regretfully, seemed out of reach.

WRONG DESIGN SEQUENCE

Designing a new aeroplane is a long and expensive process. Delays that culminated into years beyond the initial deadline are frequent, and once certified, the manufacturers will try to keep the currency of the design for a very long time. There are thousands of types of planes,[9] with some of them brand-new, flying with a type certificate issued in 1967. Such a certificate is issued by the authority confirming that a plane complies with the applicable airworthiness requirements. The seemingly dated certificates are far from uncommon; for instance, a plane that started flying in 1994 and is still in production brings an Intel-486 processor as one of its main IT elements,[10] that is, a processor discarded more than 20 years ago by many personal computer manufacturers.

Curiously, in a world where technology seems to move at lightspeed, the time since the first certification is not a good reason for a change but just the opposite. Manufacturers stay attached to the design that received the certification for decades and, paradoxically, along with crews and maintenance engineers who got their ratings to operate or maintain a specific plane become a powerful sales argument. These people will need only minor adaptations, instead of a full rating process for a new design. In turn, this relates to a huge amount of monetary capital that influences the decision about plane purchases. Hence, a type certificate is intended to last, and it is important to get it right from the beginning. Thus, the certification process must be meticulous because the design could be sustained for years.

Every system, piece, and its use and maintenance must be accepted by its merit and per the potential interactions with other elements. As this process of analysis is complex, it employs quite often the 'similarity resource', that is, asking for the acceptance of something that has been previously used in other certified planes. The process becomes even more complex when several agencies are involved in the approval of a product, especially if it is intended to be sold and fly worldwide. This can include the European Aviation Safety Agency on behalf of Europe, the Federal Aviation Administration on behalf of the United States, and the agency of the country of manufacturing. Agencies that do not automatically accept the certificate by other regulators must approve the design and, despite the existence of mutual approval agreements, a favourable decision cannot be taken as granted.

---

[9] https://simpleflying.com/boeing-737-in-service-stored/
[10] https://www.airliners.net/forum/viewtopic.php?t=757423

Large countries with huge internal markets can afford the certification of planes for internal flights (e.g., the Chinese COMAC), but this is not the general rule. To add to the complexity, regulators usually have subtle alignments with the manufacturers of their geographical areas, and some behaviours, like delegating some supervisory activities in the supervised manufacturer (Office of Inspector General, 2021), could be read in that context. This environment leads to frequent delays and increases the design costs due to the requirement of more man-hours, including more flight tests. The design-in-development stage does not directly generate revenue (like research and development costs). Some of the initial orders could be cancelled owing to delays. Even worse, the final product, if severely delayed, could be obsolete when it is market-ready owing to competition with other products. Therefore, common strategies to manage the delays are to review and streamline the design process. This can include postponing those activities that could be attended to later and focusing the design efforts on the more compelling tasks such as those that are more relevant to the certification process. This streamlining process is far from easy due to the interactions among different parts and activities of the new plane. This is the point where a choice must be made between the logical flow of the project or the demands of the different stakeholders. The mounting pressure can easily lead to conflicting priorities.

In this case, one of the many changes in the design and certification planning drove two different but related mistakes. One was not considering a key interaction, being pulled by the urgencies of the stakeholders with competing needs and adjusting the internal priorities to please them. The other is attached to a rushed decision and ignoring all the inputs that could inform its inadequacy.

Maintenance, or 'continuous airworthiness', as it is termed within the aviation industry, must be planned in the early phases of the design. It is not simply a matter of preparing the maintenance books. The design regulations impose several conditions about accessibility or potential non-voluntary interactions with other systems while maintenance is performed. Different fluid pipes, switches, and electric wiring are subject to a scrutiny of potential unplanned interactions leading to possible damage to the system or harm to the worker during maintenance or operations.

My job function was closely related to the design for maintenance, and the problems appeared at the beginning. Some of them can be hard to overcome, and the 'working in shadows' experience is shared by manufacturers because, in the beginning, the plane design is a new concept, and it does not exist yet! Hence, issues like accessibility are analysed by using models, prototypes, and simulations with CAD-CAM, instead of dealing with a finished product. In addition, the organisational environment had its own issues.

First, getting the right information was not always easy since the stakeholders were protective of their professional turf, and thus not prone to sharing information, even though the knowledge was important to the design process. Furthermore, data from the maintenance of planes involved in flight tests, which should be the perfect complement to simulation data, were difficult to get because of the internal secrecy due to cultural issues and/or fear of possible leaks to the competitors. The lack of disclosure about this valuable information was a significant obstacle to prototyping and design development.

Second, the pace of design development for maintenance was slower than expected. A crisis could arise from an announcement about delays, and there was a risk that the manufacturer would subordinate to a schedule driven by external demands rather than consider the significance of possible design-related issues affecting the final product. Therefore, the new streamlining process would not be adapted to the real necessities. Additionally, the design activity was tainted with cultural issues. Numerous meetings were taking place, during which reaching a decision, beyond setting the date for the next meeting, seemed impossible. Hence, progress was stalled. There was a silent acceptance of decisions announced by senior management levels. We were discouraged to raise further questions or highlight more design parameters since they would likely result in delays and, thus, were often disregarded.

That was the situation when the announcement came from the top about stopping a major part of the design for maintenance activities and devoting its resources to different tasks. The design process had been filled with crises, and the management actions had been similar in all of them: decreasing budgets, decreasing workforce, and reassigning resources. By repeating so in this case, a few major issues, like the access to the avionics bay, would remain unsolved. Once the crisis started, the communication link to raise such issues was severed since no one wanted to hear about it. The communication channel changed shape and form, and it became lateral rather than vertical, with the substance formed by complaints among colleagues but without any escalation to the decision-makers. An issue could be informally raised with a manager, who might listen and, very often, agree, but it would stop there. Nobody dared to raise any issue that could question the quality of an announced decision, even if that announcement seemed inadequate or ill-informed.

In some cases, delaying an activity was not a major issue since it had its own certification track, for example, avionic systems. However, in some circumstances, changes that were deemed necessary would affect parts of the design that were closed to ongoing review. Thus, these changes were unfeasible. The power to make the change was not held among those who saw the need. Of course, during the design process, analysing the solutions other competitors have adopted to address an identified problem is important. For example, even in small regional planes, it is common to have a door below the cockpit wide enough for a technician to work, and by opening it, it should be designed so that a worker could keep his/her feet on the ground but his/her head inside the bay.

Nevertheless, instead of having a door in this plane, the design included two holes to accommodate the arms of technicians. Despite one of those holes being wider to allow the technician to look inside, the access was seriously compromised. Removing or testing any piece of the equipment inside that bay was complex, if not impossible with the resources available to ramp engineers. That would mean that minor issues, easily solved on the go in other models, would require more time – and flight delays – or being postponed for them to be fixed in the hangar. In some cases, that could mean grounding the plane until the problem was fixed.

Furthermore, should external pressures dictate on-time completion of tasks and delivery of the plane for operations, engineers would probably experience severe physical discomfort imposed by the poor design, for example, leading to the adoption of poor body postures to achieve visibility or access to the bay. In combination with the physical and psychological exertion of the work requirements, repetitive similar tasks and prolonged shifts would make the perfect explosive mix of ill health and human errors for the maintenance crew.

Once the problem of inadequate working space was detected, I tried to get the decision reverted before closing the design by raising the issue with my boss and with colleagues that could be affected by the design problem. Unfortunately, consistent with the cultural norms, the issue was fully understood among peers, but no one was willing to advise a senior decision-maker to review the previous decision. A cultural environment where 'the boss is always right' produces highly disciplined organisations, and the decisions coming from above are never questioned; any mention of a potential mistake by the boss is simply unthinkable.

This single factor can explain why some cultures are excellent at getting peak performance while in known territory, but they are not able to manage agile situations where many inputs from different parts and decisions must be under constant review. The design of a new plane cannot be a straightforward process, and dealing with it as such can be very frustrating for many of the involved in that process.

At that moment of the process, any careful observer could appreciate that a comparison of the current design for a plane intended to fly in the next years with the ones of competitors already flying would not be favourable before the eyes of future buyers. However, *omertá* about wrong decisions from the top made this fact let pass or undetected until it was impossible to fix as the design of one of the affected parts, the fuselage, was to be considered final in a very short time. Therefore, this would not allow late changes that, nevertheless, would not be required, should the working space issue had been addressed.

The design of a plane is a very long process, and it can last still much longer if delays accumulate. It is a fiercely competitive marketplace, and development is occurring among other manufactures, which adds to the pressure of performance and design outcomes. Time pressures can impede a proper design process. Lacking the flexibility to make changes during the design process can mean that some identified problems remain unsolved. If this is repeated across different activities, the final product could be obsolete at delivery or simply inadequate and unsafe to operate. In his classic book, Porter (1990) showed why some nations could be successful in some markets and fail in others. The speed of decisions and room to question them in the light of new data can be influential factors. In other cases, it can be a matter of natural resources, but, very often, organisational culture can explain successes and failures.

Well before this design problem arose, someone familiar with the organisational culture of the company told me 'When they make a plan, they can go against a wall; they see it, they know that they will hit it, but they keep advancing until hitting the wall. Only then, they will change the course'. My experience from this case makes me agree with that point of view. Once they hit the wall, they will put the same energy into the new course, and for sure, they will learn, and they will design good products, as it happens in other markets. However, this learning process is slower than required, and it includes avoidable and expensive mistakes.

Certainly, there were important cultural issues in this case. The negative side of an extreme organisational discipline made it impossible to escalate major issues. Interestingly though, there was an 'organisational backdoor' that could be used by local workers, especially if they have been working for the company for a long time. This 'backdoor' had the shape of 'informal meetings' outside the office, frequently promoted by the top. Despite its informal character, everyone was supposed to attend, and it was a good place to know what is going on inside the company. Regretfully, this was a resource that could be used only by insiders to raise an issue out of formal channels. In this case, I was an outsider.

Furthermore, language problems in both directions existed. There were locals not proficient in any language other than their own and internationals who could not speak the local language. This was driving separation where the local group worries were not shared with the others. In other words, the informal way was hard to open for international staff. Therefore, even if any foreigner was present in these informal activities, it was rather meaningless and useless, especially if, during these meetings, there was no local manager well-positioned and able to understand the problem from both sides, verbally and technically.

My lesson and message from this case are that bringing external knowledge to an organisation can be harder than expected if the organisational culture imposes restrictions that render that knowledge useless. The working environment can look excellent when examining routine activities, but it can kill innovation or the feasibility of raising major issues if the culture is not conducive to change.

REFERENCES

Ballesteros, J. S.-A. (2007). *Improving Air Safety Through Organizational Learning: Consequences of a Technology-Led Model.* CRC Press LL.
Office of Inspector General. (2021). *Weaknesses in FAA's Certification and Delegation Processes Hindered Its Oversight of the 737 MAX 8.* Report No. AV2021020. U.S. Department of Transportation. https://www.oig.dot.gov/sites/default/files/FAA%20Certification%20of%20737%20MAX%20Boeing%20II%20Final%20Report%5E2-23-2021.pdf
Porter, M. E. (1990). *The Competitive Advantage of Nations.* Macmillan.

16 The Ergonomics Consultants Lot Is Not an Easy One

Rwth Stuckey
La Trobe University

Philip Meyer

CONTENTS

WORKING IN AN ADVERSARIAL CONTEXT

The setting of this case was the check-in area of a large regional airport, and the key issues concerned the design of the workplace, the work, and the management of workplace health and safety (WHS). The baggage handling at the check-in involved the workers undertaking three basic tasks: lifting and placing luggage brought to them by customers on weighing scales adjacent to the counter; lifting, carrying, or dragging each item to a conveyor at the rear of the check-in work area more than 2m away; and, finally, lifting and placing each item on the conveyor in a specific upright-and forward-facing orientation.

The hazards in the work were self-evident to the workers, with numerous injury and incident reports of musculoskeletal disorders being the most expected consequence. The work was perceived to involve actions that were repetitive, required considerable physical force and awkward postures, and were performed in cramped and obstructed conditions in a public-facing role. The worker group had more female

DOI: 10.1201/9781003349976-16

than male employees, and while it might be assumed that women would find the physical work more demanding than the men, the task execution difficulties were a common experience among all check-in personnel.

The check-in counters were provided by the employer, not the airport. At the time, the employer was a large company with a major presence in Australian air travel. The check-in counters were of a standard design and common across several airports that the company operated. Each item of passenger luggage was notionally limited to an absolute maximum of 32 kg weight. However, it was common in the airline industry (essentially for marketing reasons) that passenger bag weight limits were poorly applied, and there were no other load factors considered including baggage size, shape, rigidity, or grasp demands.

REGULATOR ACTIONS: ENTER THE ERGONOMISTS

The case was instigated by the jurisdictional WHS Regulator when they identified concerns about the number of strains and injuries incurred by check-in staff manually moving luggage at this airport. The Regulator issued a provisional improvement notice (PIN) to the employer. This required them to undertake a risk assessment of the manual handling (MH) demands and implement improved work design with the objective of reducing the risk of strains and injuries. A PIN was issued by the WHS inspector as a step in a legal process to mitigate a situation where the Regulator believed that an employer was contravening a legislative provision, and this contravention was likely to continue.

In response to the PIN, the employer hired an ergonomics consultant with an engineering background to provide the company with expert advice. This company-employed consultant ergonomist proposed a series of administrative controls that were immediate, simple, and inexpensive. The employer believed that those controls provided sufficient amelioration of the risk level and would be acceptable to the Regulator. However, in a fiercely competitive industry, the employer aimed for compliance at the least possible cost with only minor administrative tweaks, an approach their consulting ergonomist stated was adequate. In turn, the Regulator reviewed these proposed changes and declared the analysis inadequate and the mitigation insufficient, and the matter progressed to legal proceedings.

At this point, the Regulator's WHS inspector, who was managing the case, engaged us as their consultant ergonomists to obtain a second opinion. After reviewing the inspectorate's assessments and decisions, we chose to be involved, although we did not usually work with legal disputes due to their adversarial nature. However, this matter appeared to be one of a powerful organisation 'bullying' the Regulator, in a region where they were a major employer and jobs were hard to find, and we felt that we could provide useful support to the Regulator.

ERGONOMISTS AT 20 PACES

What became immediately apparent was the different 'filters' underpinning the briefings from the two adversarial organisations to their ergonomic 'experts', and the influence of the professional backgrounds of the two different consulting ergonomists on their approaches to the problem. The ergonomist engaged by the employer had an engineering background, while the ones employed by the Regulator had a health sciences background. The original discipline inevitably influences what any ergonomist understands about humans in constructed environments, including workplaces, products, and infrastructure. The ergonomist with an engineering background had extensive experience in WHS with a focus on the physical workplace, while we have qualifications in ergonomics and work from a human-centred perspective. By using a systems approach, we analyse workplace ergonomics and address physical and psychosocial factors across all system levels.

The Regulatory authority used the National Standard and Code of Practice for Manual Handling (MH COP) to assess risk, which was current at the time (NOHSC, 1990). The list of risk factors in the MH COP broadly included work and workplace design, as well as work organisational and human factors. We visited the worksite, observed the work over different shifts, took measurements, videoed actions, and obtained floor plans, baggage numbers, etc., to quantify the relevant work factors. We also spoke to workers and their managers about their perceptions of the issues and the solutions. As per the MH COP list, it became clear that various hazards and risks applied to much of this work.

Typical hazardous activities performed by the check-in staff and identified as workplace hazards, according to the terminology of the MH COP, are given as follows:

- Work activities: unable to work in an upright posture; applying force across the body, often with one hand; twisting, bending, and turning.
- Workplace layout: it forced poor postures and obstructed the pathways.
- Working postures and positions: frequent forward bending, twisting, and sideways bending of the back.
- Weights and forces: activity duration estimated to be around one hour, generally twice per day; several hundred items handled per person and shift of up to 32 kg weight, nominally; excessive force often applied to move a hard-to-grasp object.
- Characteristics of loads and equipment: varying sizes, shapes, materials, and weights.
- Work organisation: lack of sound work design, poor training of personnel, unpredictable physical demands, lack of MH aids, poor control of work times, and public-facing roles with high level of customer service demand.
- Work environment: obstructed pathways and some slippery tread surfaces.
- Skills and experience: uncertainty about what training had been provided in safe MH, and staff reported limited consultation around work design.

The risk assessment was based on the presence of certain hazards coupled with our judgement based on the evidence-based knowledge and experience and workers' reports. To quantify the hazards as far as the methods at the time allowed, we carried out a basic analysis of the postural demands using the then-version of the NIOSH lifting equation[1] to assess the physical demands on the body in the sagittal plane.

---

[1] https://www.cdc.gov/niosh/topics/ergonomics/nlecalc.html

This tool was used to provide a measure of the relative severity of those tasks that fitted within its defined parameters. As often happens, the use of the tool was limited to only part of a task being the lifting/lowering of luggage (suitcases, boxes, parcels, etc.) and movement between the lifting and placing points, the scales, and the conveyor. Therefore, other actions, such as the sustained holding, arm, trunk, and shoulder rotation used to orient items on the exit conveyor, were not able to be assessed, although they had clearly been identified as compounding the inherent risks of the work. Such limitations notwithstanding, the NIOSH equation was utilised because it had been accepted for regular use in Australian MH assessments undertaken to satisfy regulators and other legal authorities. Although developed using a US worker population, at the time of this case study (late 1990s), it was universally regarded as a legitimate tool with sufficient reliability and validity to provide useable results (Waters et al., 1994).

The fact that much of the task activity was outside the defined tool parameters (as frequently happens with the application of tools in the real world) limited the applicability of the equation. However, it was clear that the MH demands were of an unacceptably high-risk level and involved regular one-handed and asymmetrical loads. The use of the tool to assess those aspects of the work identified that when an item weighed more than 16 kg in optimal circumstances, the task was judged to be at an unacceptable level of risk. This was frequently exceeded in this workplace, where 27 kg was considered an acceptable weight in the luggage system, with bags >27 kg tagged as a risk. Notably, this 'maximum' weight was based on calculations related to the planeload capacity, rather than job demands imposed on workers.

NEGOTIATION BETWEEN LAWYERS AND ERGONOMISTS

Armed with their ergonomics experts, lawyers met to argue the case. All the ergonomic consultants agreed that there were hazards associated with the MH of passengers' luggage. However, the employer's ergonomist also agreed with the organisation's contention that less exposure in a smaller airport reduced the risk. We, the Regulator's ergonomists, disagreed, noting that the nature of the risk is unchanged because the potential consequence remains irrespective of the duration of the exposure, especially when the minimum cumulative load is considered. The employer further argued that the only real data that mattered were the record of lost time and accepted compensation claims arising from injuries suffered. This seemingly implied that the risks do not need to be addressed until a serious claims problem arose. This resembles an insurance and actuarial approach to risk management but not the approach expected by WHS legislation, which requires harm minimisation through injury risk management as practicable as possible.

Regardless, the employer's ergonomist did not challenge the more detailed and significant assessment of the level of risk and hazard identification provided by the Regulators' consultants. We presented the findings from our assessments using both the checklist of risk factors in the MH COP and the NIOSH equation tool. These assessments identified more than 30 elements of the check-in workstation and work as inadequate and of poor design. The NIOSH risk scores for those tasks demonstrated an urgent need for risk mitigation. These findings were accepted by both the Regulator and the employer's ergonomist, and it was agreed that they should be the basis for intervention, regardless of the injury claim numbers. The employer disagreed, but their objections were more a matter of form; they were unwilling to accept responsibility due to the fear of consequences that any such admission would unleash mandates for changes at other locations in their extensive system.

The approach suggested by the employer's ergonomist relied on risks being reduced if each worker used two hands and moved their feet whenever handling luggage. This administrative control relied on worker behaviour changes but, in fact, was rarely possible. It was difficult for operators to move their feet while reaching across the scale to lift luggage because the scale impeded foot placement. Using two hands was often difficult due to the types of handles commonly found (or not) on luggage and other containers. Moreover, the work design forced the work to be done laterally, across the body, often with a low lift height, contingent upon the size and nature of the item being managed on and off the scale.

The approach by the employer's ergonomist reflected a reluctance to address the issues by changing the work design and equipment. Instead, they opted for cheaper and seemingly quick fixes. Indeed, this approach is not unreasonable when an organisation has limited resources or when used as a short-term measure. However, this was a large company with the means to potentially undertake substantial improvements, particularly as these would then be applicable over time at their other airports.

WHAT HAPPENED NEXT?

The Regulator strongly opposed the administrative controls as the only form of management because our analysis had demonstrated that the work design, although typical for the specific industry, was poor, and the work was unnecessarily demanding. From this point, the actions revolved around the legal dispute process directing the Regulator, the 'adversarial' consultants, and the employer to agree on what would constitute acceptable risk control. We provided a comprehensive suite of options for risk control for consideration and, ultimately, for implementation by the employer. All proposals involved substantial improvements with a modest once-off investment that could be offset by the savings associated with costs related to injury-related absences and improvements to work productivity.

The basic problem with the system was that it functioned in two separated parallel lines, one line being the check-in counter and the other the exit conveyor. The baggage being processed was presented by the passengers to the system, then handled, and moved at and between these two points by the workers. The control options included mechanical aids to reduce lifting and holding, changes to work design (e.g., passengers lifting their luggage on and off the scales), a minor adaptation of the existing check-in counters to remove the need for twisting, the introduction of a slide/belt system between the scale and exit conveyor, buffers to position the luggage correctly on the exit conveyor, improved training, and improved risk management by the employer. As such, the proposed options were based on reconfiguring the work system within the existing structure and equipment to achieve cost-effective risk management. Four options were presented and prioritised based on cost and complexity, with option 1 the most preferred.

Option 1: Eliminate MH by the workers. This proposal positioned the passengers at a point where they presented their baggage to the workers for check-in purposes and then placed it onto the exit conveyor themselves. This virtually eliminated the MH component from the workers' duties and placed the onus on the public. This was not an unreasonable proposition because the passengers had already brought the baggage into the area for processing, and this extra task would not be repeated for each passenger. This option could be realised by using the existing exit conveyor system with minimal cost or disruption and through modifications to the existing layout as follows:

- Removal of the front counter and the feeder slides from their current positions.
- Repositioning of the front counter sections into three or four booths, each one positioned adjacent to the exit conveyor, with space between each booth for public access.
- Modification to the counter sections in these booth arrangements to provide a standard ergonomic set-up for the screen, keyboard, and related equipment and documents with appropriate security provisions including confidentiality screens.
- Realignment of the passenger queuing system so that all passengers waited in one general queue, and the first available check-in point was available to the first person in the queue.
- Positioning of the scales between the check-in point and the exit conveyor. The passenger would proceed to the check-in point, place each item of luggage onto the scales for weighing, slide the item into the adjacent holding area once tagged, and proceed with their next item. After items are tagged and tickets processed, the passenger slides their baggage onto the exit conveyor, upright and facing forward as required at present and assuming the belt remained unchanged. To assist the passengers with this action, the surface should be stainless steel or some other smooth surface or rolling system (e.g., a ball bearing-type system that would not snag wheels).
- Provision of a protective barrier on the side of the exit conveyor adjacent to each point where the passenger presents, the conveyor being open adjacent to the scale area so that the passengers can readily move their luggage into the system.

Option 2: Raising the floor height and modifying the scales and exit conveyors. This approach involved removing lifting demands from the workers to the public and reducing related postural, reaching, and bending demands. Recommended changes are as follows:

- Addition of powered feeder slides/conveyors between the scales and the exit conveyors.
- Raising the height of the scale and front counter end-of-the-feeder slides to 300 mm.
- Recessing the scale point 200 mm from the front counter to improve public access and reduce the workers' reach demands.
- Dropping the height of the exit conveyor as much as possible by reducing the size of the driver rollers and aligning the exit end-of-the-feeder slide to meet the exit conveyor, thereby creating a slope to assist luggage movement.

- Raising the floor in the workers' area behind the counter so that the feeder slides are recessed, reducing any trip hazard.
- Replacing the existing powered exit conveyor with a unit at least 700 mm width to accommodate large luggage and enable baggage turning.
- Positioning a 'trip' at the exit conveyor end of the feeder slide to turn the baggage.

Option 3: Implementation of feeder slides between the scales and the exit conveyor with one powered roller adjacent to the scale in the feeder slide to assist movement between the two handling points. However, implementing these without raising the floor surface, which was proposed in option 2, would create a trip hazard for workers.

Option 4: Gravity feeder slides between the front counter and the exit conveyor. This was the least preferred option as it still required lifting and turning of baggage by the workers and did not significantly reduce the MH demands, including reaching, twisting, and awkward postures.

Option 2, while not necessarily the preferred solution outlined in option 1, was agreed, and legal orders were made instructing these work design recommendations be implemented. The clear direction was that all controls should be adopted and implemented simultaneously as one complete system because each component was interlinked with the others. The situation would be reviewed in one month. However, unfortunately, in the end, little was achieved in the way of improvement in this workplace.

The intervention was unsuccessful because the employer implemented only one aspect of the recommended changes at any one time, rather than systemically, as directed. Since the other components were not in place, the implementation of single changes naturally failed. This slow progress appeared to be a strategy to undermine the potential of the recommendations. Actions only happened after each Regulatory review, when non-compliance was noted, and the issue returned to a hearing to seek yet further legal directives. We repeatedly reviewed the progress (or lack thereof) and advised the Regulator of the failure to implement useful interventions. Throughout, the importance of comprehensive and simultaneous implementation of all design components as an integrated strategy and as had been directed was reiterated, but to no avail.

Frustratingly, the employer proved to be a stubborn adversary to the legal WHS framework of that time. Whereas the WHS regulations were enforceable by law, their conversion into corrective actions was entirely at the bidding of this employer. As a large company and an important local employer, they chose when and how to implement the improvements. They chose which improvements were convenient to implement, and they demonstrated that the controls were not effective when implemented individually and in isolation.

THE OUTCOME AND CONCLUSIONS

Ultimately, despite repeated breaches and legal directions, nothing substantially changed. Finally, the company changed hands and within a year became defunct.

The fact that the legal interventions failed to motivate the employer to address workplace hazards could be said to represent a failure by the Regulator and us, as the consulting ergonomists. It is notable that not even the Regulator and the Court could force this company to comply. Such was the employer's influence nationally and especially in this regional town where jobs were precious. The obstructive attitude of the organisation within an adversarial legal system is the primary reason that this case was never resolved satisfactorily. Stubborn resistance to change will frustrate the best efforts to make improvements.

In the end, we did our best to provide accurate and honest interpretations of the problems and the solutions at this workplace. The fact that the ergonomists involved had somewhat differing orientations to the problems need not be a criticism. Such diversity gives the practice of ergonomics depth but may be misinterpreted by those not familiar with the discipline or not wanting to recognise the virtue of diversity. These are not issues of ergonomics alone. Diversity of opinion occurs in all professions, but in an adversarial legal system, it is often used to undermine sound and valid expertise. While ergonomics expertise to improve workplace health and safety was not able to tip the scales and persuade the employer to implement improvements this time, it very often does to the benefit of workers and employers alike (Ramos et al., 2017).

As a final point, it should be noted that modern Australian airport baggage check-in systems now include all the recommended elements for systems improvements that we had proposed in option 1 over 20 years ago in this frustrating case, including passengers undertaking their own baggage MH, lifting their luggage into a fully automated movement system in which MH by check-in personnel is eliminated.

THE VALUE OF EVALUATION

A large, long-established Australian sugarcane processing company appointed a new occupational health and safety (OHS) manager. The company operations included mills, cane trains, and railways. There were high levels of ongoing maintenance due to the corrosive and abrasive nature of the sugar product and consequently continuous work to repair and maintain heavy machinery in the mills and the railway system. The principal OHS concerns identified by the organisation were MH. The OHS manager engaged us as ergonomics/OHS consultants after proposing to management that while the company had successfully addressed the 'low-hanging fruit', the persisting MH issues were complex and required a systems approach to hazard identification, risk assessment, and risk control to improve work methods and worker engagement.

Most of this workforce was male and had been employed by the company for several years, often being the second or third generation of their family to work there. The MH activities primarily related to mill, train, and railway maintenance and were typically performed in a working environment of extreme heat and humidity. Most of this maintenance work was performed during the hotter part of the year, and sun and UV exposure was an additional problem for the railway gangers. Also, the work was noisy, dusty, gritty, steamy, sticky, smelly, and usually heavy, undertaken repeatedly in awkward and often dangerous positions.

THE TRAINING PROGRAMME

We proposed a participative approach to instil the knowledge and skills required to address the MH problems, developing teams with representatives from all workforce levels tasked to address risk mitigation (Driessen et al., 2008). This program was designed to achieve durable outcomes, with the required operational knowledge and skills remaining within the organisation and removing the need for external ergonomic interventions by the consultants. A 6-day program was developed and implemented on four occasions at different worksites over a year.

Each program was presented at a relatively central site and attended by teams from local mills consisting of operational and supervisory staff. All participants undertook or supervised the physical work, which was the focus of the MH concerns. Although they were not excluded by us, the organisation decided that senior managers would not attend the training. All participants were adults with a range of educational backgrounds; some were tradespeople, while others were skilled by dint of years of experience with the work, even if not formally qualified.

Teams from each mill or railway depot brought an identified MH issue that they wanted to be mitigated. While all the different areas did similar work, the issues brought to the training varied. The intent was that the outcomes would be shared for application across sites in the future. The syllabus was designed for adult learning and integrated accepted ergonomics and OHS principles and methods [e.g., the local Code of Practice for MH and evaluation tools such as the Snook psychometric tables, NIOSH lifting equation, and Rapid Upper Limb Assessment (RULA) (Snook & Ciriello, 1991; Waters et al., 1994; McAtamney & Corlett, 1993)], with the existing company OHS systems (e.g., job safety analyses and risk reporting and management systems).

The programme introduced the participants to practical working methods for on-site task and environmental analysis of their identified actual workplace problems. Each program culminated in presentations to middle and upper management by the course participants who supported their case studies with concise, evidence-based, costed arguments for work design improvements, prioritised implementation plans, and projected revised risk assessments for each recommendation.

From the outset, it was emphasised in the program design that there was a need to generate awareness of causative factors in the occurrence of MH hazards and develop a more sophisticated approach to risk assessment and risk control to deal with the causes, not the symptoms. The programme strongly avoided any suggestion of 'learning how to lift' as this approach focusses on changing individual behaviours, rather than eliminating the hazards at their source. Instead, the programme focussed on a systems approach to analysis and resolution, addressing all the relevant human, equipment, organisational, and environmental risk factors. The training was directed to the acquisition of practical knowledge and skills in MH risk assessment and risk control using currently accepted risk assessment methods. The use of team-based learning aimed to provide a spread of experience and ideas for improvement as well as support those with poor literacy or computer skills.

THE EVALUATION

A year after running the training, the ergonomists were contacted and advised that the company senior management had the perception that the program had not shown any evident outcomes. They asked for an evaluation of the training and its application. That was a rare and valuable opportunity for external consultants who seldom get to see what happened and why after they leave the workplace. Given that the company wanted to address more of the MH problems using in-house resources, the question that we were asked to address in the evaluation was the degree to which this had or had not been achieved and generate recommendations for improvements.

The evaluation was conducted on-site, over five days, was highly consultative and open, and included:

- Meetings with many of the program participants and OHS personnel.
- Meetings with recently appointed health, safety, and environment coordinators.
- Review of outcomes for each of the projects addressed during the training and any updated risk assessments.
- Review of any current projects being undertaken by the MH teams that had participated in the training.

Surprisingly, the evaluation determined that in fact, there had been a great deal of constructive activity by the newly trained teams. It identified an active and productive program of operational activity, including work redesign to address 138 identified hazards. Only 5% of the outcomes of these hazard management processes had not been actioned. Ongoing intervention implementation was recorded for 57% of the MH projects, and in 38% of the cases, appropriate risk mitigations were fully implemented. Most of the MH training participants had continued to undertake some MH problem-solving activities within their individual worksites, seemingly without senior management's knowledge or awareness, as explained below.

On the one hand, the evaluation showed that the participants and their direct managers were confident that the training had met their expectations. Indeed, there had been a beneficial and enduring transfer of skills and knowledge, mainly due to the support and encouragement of the project by the company at an operational level. However, on the other hand, the review identified significant gaps in resourcing and communication systems. Other than resourcing, the most significant issue was the demonstration of the success of the program to senior management.

At a high level, the sharing of successes was largely dependent on the documentation of the MH improvement activities. This, in turn, was dependent on the systems put in place to support and assist documentation of progress. The significant gap that was identified was a lack of intranet documentation, which did not allow the communication of outcomes to more senior levels of management. Consequently, senior management was ignorant of the successful ongoing and robust work design intervention activities that their workers were systematically undertaking at operational levels.

The identified gaps were further analysed to underpin recommendations for improvement; the main ones are listed as follows:

- Assigned priority to the work of the MH teams regarding time allowance for meetings, analysis and problem solving, report preparation, etc.
- Access to administrative support and computers to complete the assessments and control proformas.
- Budget provision by the management at each mill or railway.
- Engagement of the company staff OHS personnel with the MH teams to better embed the sustainability of the system.
- Recognition by management of the existence of the 'in-house' MH teams.
- Cooperation from the various internal departments at the worksites to assist with implementing ideas for improvements.
- Demonstration of practical outcomes including documentation of who was involved and how in the MH teams, and implementation of formal systems for communicating and sharing interventions and outcomes, including benefits and challenges.

Overall, the findings of the implementation and outcome evaluation of the success of the training program can be summarised by the comments offered by the participants during the programme evaluation. Most training participants reported they remained engaged in solving MH problems via the teams, but a formalisation of the MH team process and roles was required to give them official status. Nothing could happen without management commitment and allocation of resources. Through a return of investment analysis, specific MH budgets and resources had proved to be cost-effective at one work location and should be implemented at all work locations.

Also, the participants stated that problem-solving was occurring anyway, but the training offered systematic methods and tools to address the issues. Nonetheless, they acknowledged that the use of scoring methods may have legal implications and must be used with caution; while acknowledging a serious hazard exists, the risk may not be able to be adequately quantified. Moreover, documentation must be readily integrated with procedural documents (e.g., standard operating procedures), and worksheets and assessment processes can be adapted locally but must be relatively consistent across the organisation. Importantly, the total effort must be communicated and shared across the organisation via the intranet to reduce duplication and wasted effort. Evaluating the solutions is as important as solving the problems because demonstrated successes prove the worth of the program to reduce MH injuries.

THE OUTCOME

The consultants reported the results of the evaluation back to the company. The evaluation was almost entirely informed by feedback and comments from the participants and their managers. Almost all the projects being undertaken had been identified by the course participants and were being undertaken as self-directed activities. The company management expressed confidence that the training had indeed met the objectives and had resulted in a useful development of practical knowledge and skills by their workforce. We also informed the company of the existence of shortcomings in the integration of the training and its application, and lack of communications between the MH teams and senior management about the progress of the implementation of the learning to workplace problems.

OUR EXPERIENCE AND INSIGHTS

External consultants rarely can review their work and learn from how their interventions have influenced an organisation and its processes and internal resources. Very often, the handing over of a report is the last activity many of us see as the outcome of our work. Typically, unless it is wisely built into our brief, we have little or no input into the interpretation of our assessments or the application of our recommendations to achieve successful outcomes. Those who engage us to address their perception of the problem assume that the report we subsequently provide will fix things without any further change. However, most interventions need ongoing assessment and tweaking with a staged process of implementation and evaluation.

As a general principle, the value of post-implementation evaluation resides in the opportunity to review what was done against what was recommended, and whether, how, and why the aims and objectives were achieved. As consultants evaluating our own intervention, it was vital that the review identified the good, the bad, and the ugly and presented unbiased findings based on all the evidence that could be amassed. This process brings significant opportunities for learnings for ergonomics practice because it can reveal whether there has been a real or token implementation of the recommended risk controls and highlight limitations, weaknesses, and/or failures in the outcome of the intervention.

Also, the evaluation can identify limitations of the intervention methods or recommendations and provide the opportunity for a second look with the chance to define and redefine the original intentions, outcomes, etc. Furthermore, the evaluation might discover better outcomes than seemed possible at the conclusion of the original intervention. Most importantly, evaluations provide a conclusion, whereas the intervention only poses the question. Being able to achieve this is a satisfying point, regardless of the nature of the findings, which consultants seldom get to experience.

The company in this case was commendable for its initial conviction of the need for developing in-house knowledge and skills for the improvement of MH and for commissioning the training programme. Importantly, they demonstrated the commercial sense to follow up to make sure that the money that they had spent on the consultants and invested in releasing staff to attend the training was justified and productive. While it was good business sense to undertake the training in the first place, it was just as good to make sure the money had been well spent.

As consultants, we were happy to know that our work was proven to be well designed with demonstrable benefits addressing workplace MH problems and knowledge transfer, and it was favourably regarded by the participants and the company management. That we had met the brief to the satisfaction of all parties meant that although ongoing engagement with the company was largely obviated, at least as far as MH expertise was concerned, the consultants and the company parted on the best of terms, both well pleased that a useful, professional outcome had been achieved.

As consultant ergonomists and professional practitioners, regardless of where and how we are working, we are mindful of the fact that not everyone else is as convinced of the value of ergonomics as we are. Hence, it is up to us as practitioners to demonstrate good practice and be able to support our work by sound argument. As another ergonomist once opined in a conversation with the authors, ergonomics is as much an art as a science. Striking the balance between those two is the unenviable task of the practitioner. It is our responsibility as representatives of the profession to provide evidence-based and practical interventions with integrity and imagination.

REFERENCES

Driessen, M. T., Anema, J. R., Proper, K. I., Bongers, P. M., & van der Beek, A. J. (2008). Stay@ Work: Participatory Ergonomics to prevent low back and neck pain among workers: Design of a randomised controlled trial to evaluate the (cost-) effectiveness. *BMC Musculoskeletal Disorders*, *9*(1), 1–11. DOI: 10.1186/1471-2474-9-145

McAtamney, L., & Corlett, E. N. (1993). RULA: A survey method for the investigation of work-related upper limb disorders. *Applied Ergonomics*, *24*(2), 91–99. DOI: 10.1016/0003-6870(93)90080-s

National Occupational Health and Safety Commission. (1990). *National standard for manual handling and national code of practice for manual handling*. Canberra: Australian Government Publishing Service.

Ramos, D., Arezes, P., & Afonso, P. (2017). Analysis of the return on preventive measures in musculoskeletal disorders through the benefit-cost ratio: a case study in a hospital. *International Journal of Industrial Ergonomics*, *60*, 14–25. DOI: 10.1016/j. ergon.2015.11.003

Snook, S. H., & Ciriello, V. M. (1991.). The design of manual handling tasks: Revised tables of maximum acceptable weights and forces. *Ergonomics*, *34* (9), pp. 1197–1213. DOI: 10.1080/00140139108964855

Waters, T. R., Putz-Anderson, V., & Garg, A. (1994). *Applications manual for the revised NIOSH lifting equation. U.S. Department of Health and Human Services, Centers for Disease Control and Prevention, National Institute for Occupational Safety and Health, DHHS (NIOSH) Publication No. 94-110* (Revised 9/2021). DOI: 10.26616/NIO SHPUB9411Orevised092021external icon.

17 Tread Softly Because You Tread on My Dreams
Reflections on a Poorly Designed Tram Driver-Cab

Anjum Naweed
CQUniversity Australia

CONTENTS

This case is my first project involving trams and also reflects my first experiences with the political intrigue inherent in industry-focused research. From a research perspective, the insights storied here evidence a spectacular failure to include quality human factors and ergonomics (HF/E) input as a key step in good work design. Although the research findings reflect more by way of successes than "losses," there were many things I wish I *could* have done, and with the benefit of hindsight, *would* do now, were I imbued with an ability to rewind time.

Based on anecdotal comments received since, my final report from this case appears to have become a confidential and much sought-after assessment of tram design. I expected some secrecy given the political climate. If you are one of the few to have read the original report, then the "untold" experiences shared here make a good companion. What I experienced during this project left an indelible impression on me and shaped my decisions and choices in future projects.

DOI: 10.1201/9781003349976-17

I was some years into a research fellowship in Australia at this point. I had gained lots of industry experience, courtesy of leading research and being a Deputy Program Leader of the safety and security portfolio for the Corporative Research Centre for Rail Innovation. So, I had a few wins under my belt and shiny pips on my collar. I was content in semiautonomous leadership but still inexperienced in the ways of industry politics. My mind was wedded to pursuing science and blinded by its lights in ways that did not always bring the bigger picture into focus.

THE WORLD OF TRAM DRIVING

Trams are designed to connect people to cities. They operate smoothly and predictably in mixed-road traffic and through districts with high people densities. Those who drive them must navigate throngs of pedestrians who can be oblivious to traffic threats because they are not fully present in their surroundings (e.g., wearing headphones) or fixated on other goals (e.g., rushing to save time). Tram driving environments are therefore highly dynamic and challenging, but what does the driving itself entail? *"Surely all the driver does is STOP and GO?"* is a response I often hear. It is a popular viewpoint in those who know little about this occupation, including some who work *within* rail.

Trams are large, move on guided tracks, and the driver has a large field of view, so tram driving may seem easy. In practice, this job has a great deal of complexity. When operating their trams, drivers are "plugged" into their vehicle posturally, behaviourally, and cognitively, and watching a tram being operated is a skillful display of human and machine working together in complete accord.

Because trams have metal wheels and roll on metal tracks, tram handling is not straightforward. *Tribology of the wheel–rail contact* captures this in technical terms and relates to the adhesion, friction, and traction characteristics. If a driving rail wheel applies a tangential force larger than the friction coefficient, the wheel will slip. In layperson terms, this means a tram will not always pull up where you expect; if you are travelling downhill, or if it is raining, wheel slippage may occur, leading to overshoots of signals or stations, and at worst, derailment and collision. Hence, drivers must know their tram, quite intimately you might say, and develop skills allowing them to feel and drive confidently, reliably, and safely under any conditions.

The lever a tram is operated with is called a "master controller." This is the primary input that controls throttle and braking when moved in the correct direction. The degree of precision necessary to control this lever means the driver must, in effect, be a master controller ... of the master controller. It is a bionic appendage, an extension of the driver's own body. It transforms them and the tram into a *cybernetic-like* system, but it does not mean this system is error-proof. What if, for some reason, the human at the helm does not respond to a need to apply the brakes and the lever remains in the forward position?

Never fear, so-called solutions are here! Rail vehicles have systems designed with varying degrees of sophistication to detect driver state, for example, if they are distracted and slow to react, or unconscious and unable to react at all. These systems try to intervene and stop the tram with emergency brake applications. As "perception-action systems," they force a response from the driver who must act within seconds to "reset" the system, effectively to tell the machine all is okay, only for the cycle to start all over again. These systems typically take the form of a "deadman" device or "vigilance" system.

To be activated, a "deadman" device needs be held in a certain position with sustained force. Typically, it is integrated into a foot-pedal and needs to be kept down during driving, or in the master controller itself and needs to be depressed or twisted and held a certain way. Releasing the device will cause emergency brake activation. It is possible, however, for this device to be "defeated," for example, if the human collapses and their body keeps it activated.[1] In comparison, a "vigilance" system activates if there has been no input (e.g., master controller movement, use of gong) after a certain length of time. Sometimes, they may also activate in fixed intervals (e.g., every 30 seconds). In both activation cases, the driver needs to acknowledge and reset the system, with failure to do so resulting in emergency brake activation. "Deadman" devices and vigilance systems are legacy designs, meaning they have been around for a while, drivers are familiar with them, and their design issues are well documented (Naweed et al. 2020; Naweed, Bowditch, Trigg, et al. 2022).

A final point is that tram drivers are not driving for the fun of it. Their companies deliver a service, drivers receive a paycheck, and they must perform as expected. Because trams share the same tracks, drivers must avoid running early or late to prevent knock-on impacts that will leave the whole system in chaos. Timeliness is therefore a key performance indicator, and tram drivers always operate against a backdrop of time pressure. In sum, tram driving is more complex than people perceive, the environment and context of work means tram controls need to be intuitive and usable, and the cab itself needs to be a designed in a way that enables, not impedes, the multifaceted needs of drivers.

I HAVE A BAD FEELING ABOUT THIS

I was approached by an engineering firm with a real-world problem; one that required immediate attention, or so I was told. The person from the engineering firm who called me stated that this was to be a collaboration, "*a partnership!*" they announced fervently, "*we need to study a tram, and we need you to do it.*" The firm had no real familiarity with HF/E; it was a "*dark art*" to them, and for many, still is. Because of my positive reputation, they wanted to engage me under the premise that I would do the research, and they would mediate with the rest of the stakeholders.

A few units of a brand-new class of tram were in service, pending accreditation from the local transport safety authority, and approval from the rail regulator before the fleet could be commissioned. This tram was slick, with clean angular lines, vibrant colours, and when it ran, you could feel the power and performance. It sounded like a dream to all except the ones who mattered most: the tram drivers, who were in a Kafkaesque nightmare. They were complaining about the driver-cab of the tram, problems with its interface, issues with usability, and a laundry list of other concerns. The transport safety authority became concerned by the potential for

---

[1] The 2003 Waterfall rail disaster in NSW is an oft-referenced example of deadman device failure. https://www.onrsr.com.au/publications/presentations-historical-resources/waterfall-rail-accident

any habits and behaviours formed in the tram to transfer adversely to the driving of their other tram types. I started having a bad feeling about all this.

Their concerns were large enough to reach critical mass and force the involvement of impartial university researchers. With so much money invested, nonaccredited rolling stock,[2] and the regulator watching closely, there was a lot at stake. The issue needed investigation and resolution - pronto. Enter stage left, my research assistant and I, straight into the same old story: company makes product → end-user dislikes product → ergonomist must fix product.

The firm only shared some of the specifics with me, enough to whet my appetite. I started putting together an unexciting but robust study with good scope and empirically rigorous methodology. Only after we started the project did we come to realise that there was no real HF/E input in the product design in the first place.

TACKLING THE PROBLEM

I had only worked in "heavy rail" previously, an expression for passenger/freight train services. While they seem heavy to us, trams are nowhere near the size and mass of their brethren, earning them the categorisation "light rail."[3] I assumed the principles of HF/E in heavy rail would carry-over; rail was rail after all, and my methodology featured the fundamentals but also some desirables. It was a long wish list. In those days, getting sign offs for everything I wanted to do without resistance seemed a rare occurrence, so you can imagine my surprise when I encountered none. It felt like a win for me at the time. *"Gosh, this must be really serious,"* I thought to myself. In very general terms, I asked for:

- a privacy framework to collect data from tram drivers;
- focus groups with drivers;
- open consultation with broader stakeholders;
- cab walkthroughs and tram rides during full-service operation;
- freedom to assess risks/issues that may be indirectly related to the problem at hand as well as identifying the key hazards; and
- checking the effectiveness of any current control measures.

My overall approach was to ensure end-user engagement through a *participatory ergonomics* process (Wilson et al., 2005), where the workers have a say and actively engage with the design of their work environment. Table 17.1 breaks down my list into specific activities, all of which aimed to develop recommendations for redesign or more informed control measures. A lot of review work was included, with informal cab rides up front. This was to understand what had gone on before they came to me but also to gain some familiarity with the work environment.

I was also excited to apply a technique I had developed soon after completing my PhD, one I have since used widely: the *Scenario Invention Task Technique (SITT).*

---

[2] Rolling stock is a generic industry term that denotes anything on rail wheels.
[3] Trams are also known as streetcar, tramcar, trolley or trolleycar, and often referred to as "light rail vehicles" or "LRVs".

TABLE 17.1

Overview of the Activities

Documentation Review	Focus Groups (Scenario Invention Task Technique)	Drive-Cab Decomposition + Cab Observations
Review tram manufacturer HF/E documents	Tram driving tasks	Informal cab rides
Review tram mock-ups	Job design (operations, training, fleet rosters)	Cab walkthroughs • equipment • tasks • driver-machine interactions
Review company safe-working policies	Challenging work/shifts	Observe formal tram driving • equipment • tasks • driver-machine interactions
Review safety management systems guidelines from local transport safety authority	Review of new tram and other classes	-
Review driver feedback	Strategies and adaptations	

This technique draws on principles from the Critical Decision Method (Klein et al. 1989) and probes knowledge in ways enabling participants to simulate their activities in the third person and encourage deep self-reflection. The chief point here is the ability for people to generate challenging scenarios with the ability to conceptually freeze and unfreeze time so the researcher can examine why their (in)actions or (in)decisions make sense and determine what could happen when the reality changes. It was important that the cab rides occurred after the SITT, so that I could examine the scenarios concretely and validate them in the real(er) world.

SITT has turned into a valuable tool because it helps people transition from analytical and creative thinking to systems thinking. I have used it to scaffold direct methods of data collection like one-to-one interviews and focus groups, and it works especially well in the latter because participants can share their scenarios with each another and offer validation.

It has been applied in rail driving (e.g., Naweed et al. 2012; Rainbird & Naweed 2016), rail network control (e.g., Naweed 2020), aviation maintenance (e.g., Naweed & Kingshott 2019; Naweed & Kourousis 2020), maritime tourism (e.g., Pabel et al. 2020; Reynolds et al. 2021), and aged care (e.g., Naweed et al. 2021).

My methodology was approved, the project signed off, and the ethics application cleared through a low-risk pathway. The document review process was up first and very onerous. I like a bit of document analysis (see Naweed, Bowditch, Chapman, et al. 2022) but, on this occasion, I felt inclined to review all documents carefully and they were very dense and technical. The *Work as Imagined vs Work as Done* (Hollnagel 2016) concept comes to mind as a relevant summation of this experience.

I encountered lots of rule-based rigidity on what must happen in the work and idealised views of HF/E practice in cab design. It made me wonder exactly what level of end-user engagement had been undertaken.

Data collection occurred over three days and I did four focus groups with 15 drivers, with the cab observations too. It was no easy task to roster groups of drivers like this, especially given that those able to drive the new tram were in the minority. It particularly pleased me that all drivers approached to participate agreed to do so. It was a real testament to how important this issue was for them.

MY ERGONOMIC MUSINGS AND INSIGHTS

THE SEAT

The driver experiences highlighted poor anthropometric variation (i.e., fit of equipment/technical environment with different body sizes and shapes) in the physical design of the seat in the cab. Comments were like "it's too close," "the arm is not long enough," "there is no adjustment whatsoever," and "I've got sore shoulders now". Part of the issue was that the designers/manufacturers had decided to move the master controller from the console into an arm on the seat, turning it into what they called *Captain Kirk's Chair*[4] (Figure 17.1). I'm sure there was a lot of ingenuity involved, but this technical innovation robbed it of HF/E diversity. The arm of the seat could be pivoted in a downward direction at the location of the elbow in an effort to accommodate those with bigger and longer arms (most drivers), but this was reported to transfer muscular strain to the shoulder.

There were other issues with the seat, but one oddity stood out. For some reason, the relative positioning of the hand and arm used to operate the master controller favoured the throttle. Let us assume that "á" is the direction the tram is headed. This meant that pushing forward accelerated the tram and pulling it back applied the brakes (i.e., á = acceleration; â = brakes). At first, that sounds logical, but the issue is that, in other trams, it was the other way around (i.e., â = acceleration; á = brakes). The counter logic to directional mapping is that in emergency situations, it can be easier to push forward than to pull back. It therefore created transfer conflicts with other trams. Furthermore, because the arm of the seat could be pivoted downward, and this was often the observed case, the driver's hand was pronated over the master controller. In conjunction with gravity and the weight of the human limb, the design default was biased towards throttle application and speeding. This suggested that apart from insufficient HF/E input, little or no consideration was given to systems thinking of how seemingly unrelated elements can, in fact, interact and generate unanticipated outcomes.

To complement these subjective insights, I could also have used tools with quantitative measurements to better understand the physical impacts of integrating the master controller into the seat. I am not a physical ergonomist, so ensuring respective

---

[4] A pop-culture Star Trek reference to Captain James T. Kirk's command chair in the Starship Enterprise, which featured left- and right-hand controls such as Red Alert, Yellow Alert, Shuttle Operation Controls, and Intercom controls.

FIGURE 17.1 Digital illustrations of the seat, arm-rest buttons, "A-pillar" and door controls in the tram cab.

expertise was available would have been important. Nonetheless, I did not know enough about the nature of the issues when developing my methodology, and, in hindsight, it may not have been feasible from a time or cost perspective, anyway. Naturally, further ergonomic assessment to support anthropometric variation was going to be a recommendation.

THE BUTTONS

On the left arm of the seat were eight buttons, arranged in a pattern evoking divinatory geomancy (Figure 17.1). While their appearance conjured impressions of foresight, from a HF/E perspective, they violated nearly every tenet we hold sacred. Some issues with them were as follows:

- poor discrimination in button design;
- no backlighting;
- uncomfortable button action;

- quantity and clustering of buttons; and
- button accessibility.

In one previous tram class, three buttons were placed in the arm of the seat. Integrating a further five was thus a complete change in button-space and control-load mapping. The three buttons on other trams were also in slightly different places in the new tram. The buttons did not appear to have any rhyme or reason attached to either colour mapping or placement. The "Track Brake" and "Doors Lock" buttons were both red; why? The "Hazard Light" and "Sand Button" were both amber; why? The "Headlight Flasher" button was teal; why? The "Gong" and the "Horn" were both white; fair enough, but at opposite ends to one another. Drivers said to me, "yeah, I know [where the buttons are], I have an idea because we do all the wrong things," and, "it shouldn't have to be like that."

During the focus groups, I tested drivers by asking them to draw the button arrangement from memory. Some of them stole a look at what others were doing but that did not help them. Out of the 15 drivers, only one was able to do it successfully. The poor discrimination in button-design and considerations for tactility meant that drivers *looked down each time* they used them. "They don't have a little dot on the track brake button," said one driver, and "in the [other trams] the gong is more accessible than the horn," said another. The stiffness of the buttons was also a source of complaint, "in this tram, I felt my thumb was paining each time I press the doors."

A lack of backlighting made using the buttons in low-light conditions problematic. Drivers could not see the button they wanted to use, so they turned the cab light on. A great workaround, except for the luminosity flooding into the cab reflecting on the windscreen. It completely washed away the visibility of the world outside and turned it into a mirror. The driver workaround for their original workaround was to *pulse* the light on and off whenever a button was needed. This way, loss of visibility was momentary even it created a lingering retinal afterimage. So, not only were additional tasks needed (looking down at buttons, switching light on/off), but the driver's gaze and attention were momentarily taken away from the world outside. The idea that light *inside* the cab could act as a pollutant in a task formed entirely around collision avoidance was frightening and fascinating to me in equal measure.

The Sand, Track Brake, and Gong buttons are what drivers really needed easy access to. The Sand button deploys a small quantity of this granular substance onto the track to gain traction and prevent slippage, while the track brake slowed the tram magnetically.[5] In other trams, drivers flicked between these buttons with the index and middle fingers to gain smooth control, but on this new tram, using buttons was like playing a game of twister with your fingers. "How could this happen?" I asked myself. "How on earth could this design have rolled out of the production line?" Looking back now, I think this project was the first that truly shocked me, that highlighted why HF/E was as important as it was. I started thinking what they would make of our findings. "They must have some idea of these problems?" I thought to myself, "surely?"

---

[5] When the magnet on the tram is activated, it is attracted to the rail and acts on it directly, thereby decelerating the vehicle speed.

THE DOORS

The new tram also changed how the passenger doors were operated. Compared with the older trams, these changes meant that drivers easily forgot to preselect the doors on the correct side of the tram when arriving at a terminus. This increased the chances of the wrong doors being opened accidentally on the return journey with similar issues on atypically located platforms. On one cab ride, my research assistant and I witnessed the doors on the wrong side being opened. Hearing about it was one thing but *seeing* it happen was something else. We did not hear about this issue leading to an incident, but the potential for it to was very real.

Drivers deemed the ability to Force Doors Close during driving an essential feature, and used it all the time, despite company policy to use it sparingly, and judiciously. You see, the passenger doors had laser beams which automatically responded to passenger proximity, for example, by keeping doors open. Using the Force Doors Close button bypassed the lasers. This option needed to be selected prior to station arrival, but the button was sandwiched between the left and the right door pre-selection buttons (Figure 17.1). Because drivers pressed this button frequently and rapidly, incorrect door selection was more likely. Many of the issues started to feel interconnected.

EVERY SECOND COUNTS

One thing that emphasised the relevance of a systems lens for design was how little changes to timing sequences in the new tram completely unbalanced the existing system. Drivers said that the doors took longer to close, and this was verified through a cross-comparison of timing sequences during cab rides. The difference was only a few seconds, but in this world, every second counted. I remember one driver saying about the door timing, "*I feel like I'm doing overtime. I'm drained. I'm generally trying to keep time and now I'm losing time, it's stressing me out.*" In terms of prior skills, a timing delay affected the ability for the driver to provide a quality service that complied with the schedule, because it:

- Increased chances of losing right of way at intersections;
- Induced riskier decision-making due to increased delays and perceived urgency; and
- Inhibited the driver's propensity to open doors more than once, for example to allow runners to board.

The connections were becoming clearer. Frequent use of the Force Doors Close button was a strategy used to mitigate stress, so if the button was being used in a time of stress, the propensity to accidentally activate a button situated very close to it naturally increased.

HALL OF MIRRORS

Good visibility of passengers was important. Driver scenarios created using the SITT revealed that mirror-based scanning of head and crowd movements involved tacit information. In other trams, a large rear-view mirror inside the cab with side mirrors outside supported driver situation awareness of passenger movements. The rear-view mirror was much smaller in the new tram, and the side mirrors had gone, replaced by cameras feeding video into panel displays drivers needed to access through complicated menus. *"There is only one mirror, and it is too small,"* and *"there's no mirror, other than that one there…"* were the sort of retorts I received when quizzing drivers on their mirror situation.

For some drivers, the new rear-view mirror was so unusable they omitted it from their tasks altogether. Drivers were resistant to using cameras to divulge passenger movements because of the time and task load involved, and because it lacked the fidelity of a mirror image and provided *"unreliable"* and *"second hand"* information. More guesswork for door open and close timing was therefore used above observing of actual passenger dynamics. In conjunction with the Force Door Close activation, this invariably increased the risk of slamming passengers and their belongings (e.g., strollers) between doors. Admittedly, the cameras were not all bad. The increased length of the new tram meant side mirrors may not have facilitated viewing down the length of the tram. But when driving in wet weather, video images were blurred by raindrops, and, during the night, light pollution from road vehicles contaminated the picture. The concern was that cameras provided a warped and distorted representation of reality.

THE PILLARS

Finally, a main area of concern was an "A-Pillar" design of the left and right sides of the cab. This meant drivers viewed the outside through two thick pillars that angled towards the top in an "A" shape. The size of these pillars created a visual obstruction or *"blind spot"* for drivers. During observations, I saw it was possible for people to be completely concealed by the "A-Pillars". What made it worse was that a box-control panel has been situated *between* the "A-Pillar" on the left-side (Figure 17.1). This rendered smaller people, like school children, all but invisible. A poor design indeed in a country where driving occurs on the left! Drivers regularly leaned out of their chair to look around the pillar. It amazed me that feedback from drivers was needed to raise this as an issue. *"Surely, they would have known that the A-pillar and placement of the box obstructed viewing?"*. I cannot believe that any HF/E input or consultation with drivers went into this at the correct stage in asset design.

MY "DAMNING" DENOUEMENT

The transcription and analysis took a little longer than planned. During report-writing, I paid attention to detail. The project background and methodology only rang up seven out of 35 pages, so most of the report was the findings. I emailed my first draft to my contact at the engineering firm and then we spoke over the phone. *"The findings are pretty damning"* my contact said. *"Really,"* I thought to myself, waiting for feedback on what needed to be tempered, clarified, adjusted, or explained with better context. There was none of this. Proof-reading of my report or marked up comments would not be forthcoming. We met in person not long after. *"The findings are pretty damning,"* my contact said again, quite matter-of-factly without meeting my gaze.

What did it mean? Well, of course, I knew what the word "damning" meant, but what was getting lost on me was *why* my findings were being described this way, out of all the adjectives available. The word suggested extreme criticism and implied guilt or error. It also hinted I was providing a testimonial of some kind, which could lead to condemnation or ruin. *"It's not a testimony"* I thought, *"and no one is being blamed."* It is a symptom of whatever is driving the design of the system. All it meant is that HF/E input, or in-depth consultation with HF/E practitioners, had not happened. That was hardly a surprising outcome. From my perspective, the report identified what literature supported, and we were simply connecting the dots and showing them what had been missed.

In my pursuit to document, capture, and share all problems as comprehensively as possible, I did not appreciate the perception of others, nor the massive time and cost implications my recommendations would have (e.g., retrofitting). Had I seen the bigger picture? Maybe not. Or maybe I had seen it but did not recognise it for what it was. My contact had already sent the report to contacts at the company, and a full stakeholder meeting was being arranged where findings would be presented. I was told I would be the one leading the presentation and sharing the results with all present.

THE PRESENTATION

I put together slides of all my findings. There were plenty of photos. More than half a dozen people were in the room with me. I was sitting at the end of a rectangular table, closest to the screen where the laptop cabling was located. It was not a comfortable position. I had to present seated, crane my head upwards to see the slides and then turn my gaze in a ~130° motion to the left if I wanted to see the faces around me. At the farthest end of the table on the same side was someone who represented the company; I believe they had played a role in the procurement or design process of the tram. Let's call this person Charlie? Directly opposite Charlie were two people representing the rail regulator. Was I nervous? A little. The "damning" comment had stayed with me, but I was also excited. I still believed my role was to share my findings as frankly and impartially as possible. Remember that my contact, who sat opposite me, had said that my report had been sent to the company. Everyone's demeanor seemed positive. I felt comforted by this, assuming they had already read the report.

During my presentation, I provided a background of the project before describing tram driving from a system perspective. I wanted to create common ground, so everyone was aware of things that could influence performance. I shared information on methodology, sample data from scenarios, and non-specific points about safety-critical tasks and the relevance of skill transference. I had 51 slides, and my findings started on slide 17. They were organised in 11 sections and prefaced under the heading: *Pathways for Addressing Identified Issues*.

I described everything I have shared with you. Their faces were inscrutable. Till then, everyone had remained silent, and I assumed it would be like this for the duration, like being at a conference. After covering the master controller, I moved on to the buttons, calling up a photo along with the list of the issues. The silence from the audience broke. *"Err, I think you'll find..."* someone said. I craned my head to the far end. Let me tell you something, Charlie was open for business.

Charlie mounted a defence for the integrity of the button design on the basis they had formed and consulted on this with a small reference group of drivers. They had indicated which button options they liked, shared aspirations to have controls at their fingertips, and I think they even created drawings of the layout (perhaps one or more were into geomancy?!). Charlie displayed complete conviction, arguing that they had done the best possible job. I was quite taken aback by this revelation and the force of Charlie's assertions. I guess I did what any researcher would have done in my place; I started citing research. I referred Charlie to Don Norman's design principles, illustrating how so many tenets of good design were *"violated"* by the buttons. It was a strong comeback. Too strong, maybe. I don't think it turned into an argument, although it felt that way. Then I said something in exasperation that I remember distinctly to this day. I said, *"the problem is you have consulted your drivers not only as subject matter experts of tram driving, but as experts in HF/E design. They are expert at how to drive trams, not at HF/E science."* There was silence. It looked like a penny had dropped. The matter on the buttons was closed.

I continued, moving from one finding to the next. Charlie kept pace with me, trying to explain their logic of each design. It went on like this. My memory is foggy here, but we got to a point where the regulator stepped in. It was clear they understood the implications of my findings. I do not recall getting through every finding, but I do recall looking up at my recommendations slide. A final exchange from the Regulator to Charlie went something like, *"these are Anjum's findings and recommendations. You need to address each of these in turn. Do that so that we are satisfied that the tram is safe, and we will sign it off."* That was the end of the meeting. The presentation was over.

MY REFLECTIONS

At the time, I remember feeling attacked by Charlie, and validated (perhaps even *protected*) by the regulator. I was an early-career researcher, and I still stand by my findings, but at the presentation, I expect Charlie felt every bit as attacked as I did. The manner in which I delivered my findings was probably distressing and accusatory, and I understand why it would have induced a defensive reaction. I am reminded of Yeats' final line of poetry from Aedh Wishes for the Cloths of Heaven: *"Tread softly because you tread on my dreams."* Hindsight is a double-edged thing, especially when you are the one doing all the treading. Of course, my contact had been right, the findings were "damning", and I had put together my presentation too much like my report. All the time, effort, blood and sweat pumped into making these trams were at stake. And here I was, a young upstart researcher, trashing the intentions of the many who had been involved. In hindsight, I now see it was my first real foray into the politics; all that fieldwork, and I had not once thought to temper the findings by considering their perspectives. My own contact had been little more than a litmus test.

A month after my presentation, I delivered a second one, this time to the senior leadership team of the same tram company. It was a similar set up but a bigger room. Many were in the audience, all representing different departments. I was more mindful of my delivery this time, but I do not think that mattered. There was no emotion from the audience, just an interest in what I found and what I believed should be done. They wanted to know about the findings as plainly as possible. A damage limitation or rather, "damage correction" exercise that they needed to be across.

If I could rewind time, I would put more care and consideration into how I communicated my findings in that first presentation. To do this means acquiring information I was not privy to and had not thought to seek. It was only at the presentation that the existence of a reference group of tram drivers had been confirmed. I believe such a group was formed, but I question the degree to which they were involved. Had they seen the seat? Had they understood the ergonomics of the master controller? Did they know about the new sizing of the mirror? Were they aware of the extra timing for door closure, or the changes to the ways doors were operated? Doubtful. They would certainly have shared strident concerns about the "A-Pillar" design before its unveiling. I would like to have consulted Charlie before the meeting, too. Because of the nature of the assessment and the engineering firm functioning as the intermediary, my perception was that I needed to keep distance. My approach since is always to talk to the various company strata before getting stuck in the work.

A key insight I gained from this project was that light rail is not heavy rail, and the two modes could not be more different. In heavy rail, you are not penalised for running early, but in light rail, being early is worse than being late. Both modes have their challenges, but my appreciation for tram driving only grew. Trams had a unique culture all of their own. I think the misplaced assumption that the two modes are generalisable runs deep, and one reason heavy rail standards are used to inform light rail standards from the top-down.

I still support that participatory ergonomics are the bedrock of applied HF/E. My proclamation that end-users with task expertise are *not* the same as practitioners with HF/E expertise was an important self-realisation. Sadly, I also think a lack of this distinction is common in industry and one reason organisations find themselves in a pickle. Insights from end-users are invaluable, but they need to go through a well-informed "filter". Separating the wheat from the chaff may be one of the roles of HF/E practitioners.

When writing this chapter, I investigated what changes were made to the tram, and I discovered that the A-Pillar issue was taken seriously. An award-winning engineering solution was used to "fix it" in the form of analogue cameras outside the tram and customisable screens on the inside where the "A-Pillars" were. I am happy to have facilitated the success of others, but it was an expensive solution that may have been easily remedied with proper HF/E input in good work design. For better or worse, the tram cab is seemingly evolving in the direction of the glass cockpit. I am glad that our research was a step towards achieving system safety; it makes it all worthwhile.

Last but not least, I gratefully acknowledge the research assistance from Ganesh Balakrishnan, my partner in crime who experienced much of this story with me first-hand.

REFERENCES

Hollnagel, E. (2016). The nitty-gritty of human factors. In S. Shorrock & C. Wiliams (Eds.), *Human factors and ergonomics in practice: Improving system performance and human well-being in the real world* (pp. 45–64). Boca Raton, FL: CRC Press.

Klein, G. A., Calderwood, R., & MacGregor, D. (1989). Critical decision method for eliciting knowledge. *IEEE Transactions on Systems, Man, and Cybernetics, 19*(3), 462–472.

Naweed, A. (2020). Getting mixed signals: Connotations of teamwork as performance shaping factors in network controller and rail driver relationship dynamics. *Applied Ergonomics, 82*, 102976.

Naweed, A., & Kingshott, K. (2019). Flying off the handle: Affective influences on decision making and action tendencies in real-world aircraft maintenance engineering scenarios. *Journal of Cognitive Engineering and Decision Making, 13*(2), 81–101.

Naweed, A., & Kourousis, K. I. (2020). Winging it: Key issues and perceptions around regulation and practice of aircraft maintenance in Australian General Aviation. *Aerospace, 7*(6), 84.

Naweed, A., Balakrishnan, G., Bearman, C., Dorrian, J., & Dawson, D. (2012). Scaling generative scaffolds towards train driving expertise. In M. Anderson (Ed.), *Contemporary ergonomics and human factors 2012: Proceedings of the International Conference on Ergonomics & Human Factors 2012* (p. 235). Blackpool, UK: CRC Press.

Naweed, A., Bowditch, L., Chapman, J., Dorrian, J., & Balfe, N. (2022). On good form? Analysis of rail signal passed at danger pro formas and the extent to which they capture systems influences following incidents. *Safety Science, 151*, 105726.

Naweed, A., Bowditch, L., Trigg, J., & Unsworth, C. (2020). Out on a limb: Applying the person-environment-occupation-performance model to examine injury-linked factors among light rail drivers. *Safety Science, 127*, 104696.

Naweed, A., Bowditch, L., Trigg, J., & Unsworth, C. (2022). Injury by design: A thematic networks and system dynamics analysis of work-related musculoskeletal disorders in tram drivers. *Applied Ergonomics, 100*, 103644.

Naweed, A., Stahlut, J., & O'Keeffe, V. (2021). The essence of care: Versatility as an adaptive response to challenges in the delivery of quality aged care by personal care attendants. *Human Factors, 64*(1), 109–125. DOI: 10.1177/00187208211010962.00187208211010962.

Pabel, A., Naweed, A., Ferguson, S. A., & Reynolds, A. (2020). Crack a smile: The causes and consequences of emotional labour dysregulation in Australian reef tourism. *Current Issues in Tourism, 23*(13), 1598–1612.

Rainbird, S., & Naweed, A. (2016). Signs of respect: Embodying the train driver-signal relationship to avoid rail disasters. *Applied Mobilities, 2*(1), 50–66.

Reynolds, A. C., Pabel, A., Ferguson, S. A., & Naweed, A. (2021). Causes and consequences of sleep loss and fatigue: The worker perspective in the coral reef tourism industry. *Annals of Tourism Research, 88*, 103160.

Wilson, J. R., Haines, H. & Morris, W. (2005). Participatory ergonomics. In J. R. Wilson, & N. Corlett (Eds.), *Evaluation of Human Work* (3rd ed., pp. 933–962). London: CRC Press.

18 Creating Conditions for Successful Design-in-Use

Lidiane Narimoto
EWI Works Inc.

CONTENTS

Brazil is the biggest producer and exporter of sugarcane, reported supplying 50% of the world's sugar (USDA, 2021). Until 2007, 70% of the total production was harvested manually (CTC, 2012), with workers using a sharp knife in a very labour-intensive activity (Alves, 2006; Vilela et al., 2015). It all changed that year, when an environmental agreement was signed, forbidding burning prior to harvesting, hence the mechanization of the process.

I started my Master's in Production Engineering focused on Ergonomics in 2010, when Brazil was in the middle of this transition from manual to mechanical harvesting. It is estimated that the country's mechanisation index now varies from 72% to 97%, depending on the region (CONAB, 2021). The operation of such complex and expensive machinery required new job roles in Brazilian fields, and my research focused on the ergonomic analysis of the tasks. What was it like to harvest sugarcane with an enormous machine instead of a knife?

During my master's research, I was spending days at sugarcane crops, assessing the work and discovering its characteristics and constraints. I analysed the operations in the most varied and diverse situations, including tangled sugarcane stems, foggy weather, night, and sloping grounds. Harvesting erect sugarcane in flat grounds on a sunny day differs completely from harvesting under adverse conditions. It was during adversity that the equipment's limitations and design flaws became evident to me. I also analysed the work during the off-season period, where the harvesting teams work at the mill's maintenance building for three to four months. Together, they disassemble the machines, wash and clean all parts, make repairs, and assemble machinery to be ready for the next season.

However, during my research, several other interesting observations emerged. During the off-season period, besides the required maintenance, the teams did more! They made design improvements to the machinery. For example, when working with the operators inside the machine's cabin, it was common to hear them say that the

DOI: 10.1201/9781003349976-18

team had changed something in the machine that produced an improvement. I was curious about this. What also intrigued me was the fact that no brand-new machine was put directly in the field to harvest. Instead, it first had to go to the maintenance building for preparation. It was like buying a new car, not driving it immediately, and sending it first to a mechanic. How could state-of-the-art, expensive capital equipment be delivered in a state that was not ready for use?

When I finished my master's, I felt like I had many more questions to answer. Thus, I started my PhD aiming to address some questions revolving around a central topic: the design of sugarcane harvesters. I wanted to analyse the design flaws that I had seen in the field and study the improvements the operators had so proudly boasted about for the past two years. Because the machines were originally invented in Australia, I wondered about their operation in their place of origin. Was it the same and, if not, why not?

Parallel to my academic research, I was a consultant in some projects that the research laboratory, of which I was part, operated in cooperation with industries. I divided my working time between agricultural work and industrial work. In industry, it was common to see workers making minor modifications to improve their work conditions to achieve productive outcomes. In the review of scientific literature, the concept of "design in use" is not new (Rabardel, 1995; Folcher, 2003; Beguin, 2008; Rabardel & Beguin, 2005). Research has showed that design evolves through time and that users constantly modify artifacts or attribute new uses for them. What was not clear to me was how they did that. Operationally speaking, how do workers, who are not qualified designers, design? Therefore, I used my PhD as an opportunity to investigate the matter in depth.

THE AGRICULTURE CASE

In Brazil, the sugarcane harvesting team is composed of the harvester operators, tractor drivers, mechanical technicians, truck drivers, and the team leader/supervisor. Outsourcing the harvesting is not a common practice. Generally, the team works for the sugar mills that also own all the equipment. During the harvesting season, operations are 24×7. Mechanical technicians are in the field to ensure minimal operational downtime with the harvesters being repaired onsite. Repairs are mostly done in the fields and taking the machine back to the maintenance building is the last resort. During the off-season, all machinery is transferred to the building, and the team works as assistants to the mechanical technicians leading the maintenance plan.

During my research, I identified over 50 design modifications made by the on-site mechanical teams. I classified them into three groups:

- Structural modifications: reinforcements/replacements of the machines' structure in sections that would crack because of harvesting in adverse environmental conditions.
- Functional modifications: solutions to respond to design flaws and specific problems encountered by the teams.
- Operational modifications: solutions aimed at operational improvements. These included innovations of the teams beyond responding to design

problems by adding and modifying equipment features. For example, the teams installed additional lights at the back of the machines to be used during the harvesting at night.

Therefore, design-in-use occurred in reaction to design problems but also proactively to improve different aspects, such as reliability, operation, productivity, safety, and maintenance. In another example, they installed a water container that reused the water from the air conditioning system so that operators could wash their hands after their maintenance tasks. Such approaches proved so efficient that the engineers and manufacturers' representatives often visited the fields to catalogue the corrections and improvements for retrospective analysis, incorporating the design ideas in future machines. Hence, design-in-use practices became part of the ongoing quality improvement process.

When I analysed how the users designed and why their initiatives worked, two main determinants were found, namely, collective work and formal support. Collective work is related to the cooperation among different workers and the combination of their expertise. On the one side, there is an operator who excels at controlling and operating the machine in the most diverse situations. On the other side, there is the mechanical technician who has a rich background and experience in fixing, assembling, and disassembling various types and models of harvesting machines throughout the years. Then, these two actors are placed together in the field and face daily challenges. The operator notices a problem and reports it to the technician, who is seen as an inventor of solutions. Together, they elaborate and discuss alternatives. A process of brainstorming begins. They suggest ideas to one another, verify their feasibility, and anticipate possible outcomes. A solution is agreed, then implemented in one machine by the technician and tested by the operator during harvesting. Adjustments and corrections can be made and, if proven effective, the solution is replicated across the rest of the machine fleet.

However, while the combination of different knowledge is crucial, is not enough. The formal support by the organisation is necessary to allow innovation. The sector is driven by continuous improvements, ranging from cultivating resistant and productive plant species to new planting techniques and logistics strategies. Improvements in the harvesting processes are especially critical because they determine the quality of raw material obtained and the production for the next seasons. Therefore, sugar mills welcome actions that lead to increases in productivity, reductions of losses and promote cost-saving ideas. Indeed, this is not a novelty in organisations operating in capitalistic environments; however, there are different approaches to effectively employing the concept.

The work of the harvesting team is structured in a way that allows the development of social spaces and exchange of ideas and best practices among the workers. Flexible structures provide for the autonomy of workers to put their ideas into practice and, finally, the means for their realisation in terms of equipment, tools, and financial resources. In other words, the harvesting and the maintenance process are organised in a fashion that allows workers to interact with each other and to make, buy, develop, and test solutions.

For example, there are two top competing manufacturers of sugarcane harvesters in the world. It is possible to spot their machines in the sugarcane fields by their trademark colours of green or red. The mills prefer one manufacturer or the other to maintain a homogeneous fleet. This allows better planning and efficiency (e.g., training and skills in the operation and maintenance of one harvester type, supply of parts). Thus, I was surprised when I saw a green component in the maintenance building of only red machines. As the workers were frequently experiencing problems with the bearings, they were given the resources to purchase a bearing from the other manufacturer just to compare and practice reverse engineering (i.e., deductive reasoning to learn how a product works).

Obviously, workers' innovations have limits. They design with what they are given, usually an already designed product that does not necessarily account for their required tasks and capabilities, and with what else they have available. They must use the materials within their proximal environment, and they literally produce bricolages, meaning novel constructions from diverse and unusual supplies. Nonetheless, some problems needed a complete redesign initiative to achieve resolution. This can only be performed by the engineering department of the equipment manufacturers.

During my research, I also contacted the two manufacturers to understand a little about their design process and whether they had implemented participative, codesign approaches with workers. Unfortunately, neither shared details about their processes, rationalising that such information was sensitive and proprietary. Nevertheless, the structural modifications emerging as necessary during operations were witnesses that original designs did not entirely consider the task demands of the users in the often-harsh harvesting conditions of Brazil.

THREE EXAMPLES FROM MANUFACTURING

Contrasted with farms, industries such as manufacturing offer more controlled and standardised settings. Throughout my years of practice as a consultant, I noticed that workers' bricolages in various industries are usually perceived by engineers, supervisors, and management in general as failures or aberrances to prescribed work, or an act in defiance of the rules. Organisational structures usually separate the different functions into departments with specialised roles (e.g., process engineers, quality improvement and maintenance). In the manufacturing industry, this departmentalisation is more noticeable, and it is not expected that workers do a job falling outside the sphere of their responsibility. However, this fragmentation of the work design separates the technical sectors from the workers.

In Portuguese, there is a word called *gambiarra*, an informal term to designate an improvisation. It usually has a funny or pejorative connotation of a contraption, or a precarious, sloppy idea. More recently, a positive connotation has been identified, indicating a clever way to overcome adversities (Boufleur, 2006). At the shop floor, it is common to hear from management that whatever solution made by workers is a *gambiarra*, with a touch of disregard. Having in mind only the work as prescribed and unaware of the work as done (Shorrock, 2016), people with high-level managerial roles are usually not aware of workers' solutions and, when they become aware, there is a tendency to see these solutions as unnecessary and even potentially dangerous.

To illustrate the above, below I am describing three examples from the time I was a consultant. Two cases are from a company that produces medium density fiberboards (MDF panels) and one from a company producing corrugated pipes in high-density polyethylene (HDPE). My role in these cases was to conduct ergonomic assessments and deliver reports with the necessary recommendations to the clients.

The first example is about a job that involved the supervision of a process that occurred in large tanks containing the liquids used in the process. Because one tank did not have an indication of its level, the worker used the following strategy: carry and position a ladder next to the tank, climb the ladder, use a hammer to hit the tank in different locations, listen to the sound produced, and climb down. Based on the sound, the worker had an idea of the product's level inside the tank.

From the workers' perspective, it was clear they knew the solution to their problem; they simply did not have the means to implement it themselves. *We have been requesting an improvement for many years now. They say they must place a sensor, but a float device like a ball would work; it is cheap, and it can show us if the tank is full or not.* The exact liquid level would be shown on a scale linked to the float device by pulleys, a simple solution with significantly positive impact. Similar simple devices of float switches are used to alert blind individuals to fill mugs of beverages without spilling anything, especially scalding hot liquids. In this case, the device emits an audible alarm and vibrates when liquid levels get close to the overflow threshold inside containers.

Workers had been doing the hammer strategy for many years, and the hit marks on that specific side of the tank were noticeable by a careful observer. For safety reasons (e.g., risk of falls), the superiors insisted workers should not use the ladder. But without having a level indication, how would the workers acquire the necessary information? Climbing the ladder to hammer the tank was their only option, and the organisation would look the other way while they did it. This was an informally accepted aberrance from their written work procedures. My impression was that the workers were resigned as they had no expectations the organisation would implement the float device. They mentioned it was stressful to work under conflicting requests of not climbing while not having the information necessary to perform their tasks. Upon reflection, I wondered "What if there were an accident and the worker fell from the ladder"? Would it be classified as an "unsafe act," deferring to the behaviour of the employee, because "the worker was not supposed to climb there"?

The second example regards a job involving the visual inspection of impregnated paper sheets. The machine cuts the sheets and piles them at the end of the production line, where two workers are positioned. They inspect each sheet, looking for stains, folds, insects, dirt, print failures, etc. The rejected sheets are removed from the pile, and the approved batch is transported to a different production line that places the sheets on the MDF panels.

For the paper sheets to slide over the pile, it was necessary to reduce their static electricity; otherwise, the sheets stuck together and caused paper jams. It was like the jams inside of a copier, except that the length of industrial paper sheets ranges from 2.5 to 5.5 m. This was achieved through the application of an electric current or electrical discharge. The machine had two fixed bars for that purpose, but they were not effective in eliminating the static electricity from the papers. Thus, additional metal plates were provided by the engineering department to serve as supplementary sources of electric current. Depending on the size of the sheet produced, the workers placed three to five additional metal plates on each side of the pile, left and right. Because of the limited space, the metal plates' handlers had contact with the machine's guardrails. Workers did not use gloves due to the nature of the work that supposedly required only visual inspection without manual handling. Consequently, workers were frequently subject to electrical shocks.

The solution devised by the workers involved placing rubber wrapped in tape all around the guardrails. This was an insightful adaptation to overcome two design problems. First, the original fixed bars that did not work properly in the first place, and second, the exposure of the workers to not lethal, still uncomfortable levels of electricity. When I saw the rubber wrapped in tape and asked about it, workers responded: "oh this is a *gambiarra* that we had to create ourselves, never mind."

The organisation was aware of the situation because the workers talked to their immediate supervisors about the issue. However, the organisation's perspective was that the matter was not a concern anymore because the workers had temporarily insulated the guardrails. The workers felt unheard because the ideal solution to correct the original fixed bars was out of the question for the organisation. This solution would eliminate the need to position manually the additional metal plates and the risk of exposure to electrical shocks and would also improve the quality of the paper. Was there a better way to isolate the guardrails? In the field, some workers' ideas are clever solutions that often just needed either to be recognised and accepted or slightly improved and better developed before implementation. An appropriate insulation of the guardrails should have been provided by the organisation. It would be as functional as the rubber wrapped in tape, but it would make the impression that the organisation addressed the problem correctly. What if the workers' solution was not adequate according to safety standards? Would the organisation seek strategies that were compliant?

The third example is from a job involving the fabrication of special connections of corrugated pipes. A corrugated pipe has a series of ridges and grooves running parallel to each other on its surface. The ridges and grooves follow a pattern that is perpendicular to and bisecting the centreline of the pipe. Corrugated pipes are made of different materials, such as iron, steel, polyvinyl chloride, or HDPE. Corrugated pipes are used where flexibility is an important factor other than strength and durability, such as in storm drains and culverts. Flexibility makes corrugated pipes more useful and suitable for a wide variety of uses compared to rigid and noncorrugated pipes.

In the company, large plastic extruder machines produce such pipes of diameters ranging from 8 to 60 inches. The job of interest belongs to a series of customised solutions based on clients' requests, for example T-joints and L-joints. To make the joints, the workers must cut the pipes with an automatic or manual saw, depending on their size, and glue the pieces together. This bonding is made using an extrusion welding gun, which heats the plastic and extrudes it as a molten material. Like metal inert gas welding for metals, the weld beads of plastic joint the pieces of pipes. The weld beads are smoothed and finished while still warm with tools that knead and scrape. Those are unique tools that are not commercialised, without technical names; both the "scraper" and the "kneader" are made by the workers. To build the tools, they use the extrusion welding gun to make the handle, and, at one extremity, they shape it like a pestle of a spice crusher (in case of the kneader) or attach the metal blade (in case of the scraper). They make various tools of different sizes, and the scrapers are made with different types of blades. Workers were proud of their creations, walking around the sector looking for every tool that they ever made to show me.

In this example, there was a positive facilitator. The same tools used to weld also allowed workers to shape anything they needed in plastic, just like playing with modelling clay or a hot glue gun. The organisation provided the blades, and the workers made the handles. But what if there was no extrusion welding gun? Would the workers have to adapt to use existing off-the-shelf tools that are available for purchase? Would the maintenance sector intervene? Or, most unlikely, would the organisation find a supplier that could design customised tools?

LESSONS LEARNED

In the case and examples described above, I had different roles as a researcher and consultant; still, on all occasions, I was a work analyst. My goal was to unravel the work as done and understand the activity in the light of work constraints. When you are in the field analysing work, it is almost impossible not to come across any kind of workers' bricolage, *gambiarra*, design-in-use, or any other term of choice like "work-around" in several countries. They exist even when the work is properly designed, let alone when it is not!

Workers design with what they have and what they can find in the environment by combining, recombining, and reorganising available materials. My learnings are that workers' design initiatives appear first as a response to problems and constraints that they face daily when trying to complete their tasks. Problems can increase workload and/or impact productivity. This was true in the sugarcane harvesting case when workers fixed structural and functional problems of the machines that impacted operations or even stopped the harvesting activities.

The same was also true in the example from manufacturing when workers used a hammer and a ladder to learn about the tank's level and wrapped rubber to protect themselves from exposure to electricity. The workers' attempts to address and solve the problems they experience may be effective or partially effective. Cases when such strategies are only marginally effective still offer insights to work analysts into possible paths to be pursued. I also learned that workers' inventiveness does not stay in the sphere of existing problems. They tend to go beyond and show highly innovative skills when the right conditions are provided, such as when they can interact with others to exchange and combine experiences and have the technical means or resources.

In the manufacturing case of special connections of pipes, I believe it was a fortunate coincidence that the tool designated for welding allowed creating other tools the workers needed. On one side, there was a problem of not having the right tools to knead and scrape and, on the other side, there was an extrusion welding gun which allowed them to craft anything in plastic. It might have been that the organisation did not even know about it at senior management levels.

In the example from the sugarcane harvesting, conditions involved more than the technical means. The teams were proactive and innovative because, out of necessity, the organisation developed a favourable culture which was supportive of worker design initiatives. Workers were given agency, space, equipment, and financial support to test their ideas, even if they were just a hypothesis. Most importantly, the organisation tolerated some degree of experimentation and failure, which are necessary preconditions for iterative designs.

Clearly, Brazilian sugar mills are an exceptional demonstration of users' design at its optimum. I do not suggest that organisations should do the same. However, they must do the minimum which is to consult with their workers. This is already foreseen in work health and safety legislation in many countries, but consultation does not have to be limited to health and safety. It can embed all work aspects, especially considering that health, safety, productivity, quality, and other organisational objectives interrelate and together contribute to business success.

Experienced workers know more about their work than anyone else, and it is only logical that they must be allowed to think and act on work (re)design. The idea of them testing a solution previously selected and imposed on them by others is not enough. It is necessary to include workers during the entire process. Such an inclusive approach generates effective solutions and engages those who matter most, the people who do the work. Moreover, people tend to experience meaning in their work when they feel they can contribute to creating something of value, especially when they feel able to explore, connect, and create a positive impact. Organisations that recognise the benefits of employee involvement can improve performance, productivity, safety, job morale, health profiles, socialisation, and overall workplace culture.

During my research and industry activities, my failures and frustrations were felt when my communication fell flat. The ideas of workers and mine were not appreciated, and any resolution was never formally acknowledged. More alarmingly, in those instances, the workers had to hide their good work out of fear that its discovery or contribution to something "wrong" could reveal modified approaches not formally approved by management. On the other hand, I felt successful mostly when I communicated the findings to a receptive organisation that was prepared to track these changes and recognise the inventive nature of their employees through continual design improvements.

REFERENCES

Alves, F. (2006). Por que morrem os cortadores de cana? *Saúde e Sociedade* 15(3): 90–98.
Beguin, P. (2008). Argumentos para uma abordagem dialógica da inovação. *Laboreal* 4(2): 72–82.
Boufleur, R. (2006). A questão da gambiarra: formas alternativas de desenvolver artefatos e suas relações com o design de produtos. Masters Thesis. FAU-USP: São Paulo.
CONAB. (2021). Acompanhamento da safra brasileira de cana-de-açúcar. Safra 2021–2022. *Companhia Nacional de Abastecimento. Brasília, Terceiro levantamento* 8(3): 1–63. Available at: https://www.conab.gov.br/info-agro/safras/cana/boletim-da-safra-de-cana-de-acucar/item/download/39836_dace9b05e78210b93d898b3ff45f19c8
CTC. (2012). *Censo varietal e de produtividade em 2012. Centro de Tecnologia Canavieira.* Available at: https://docplayer.com.br/15315609-Censo-varietal-e-de-produtividade-em-2012.html

Folcher, V. (2003). Appropriating artifacts as instruments: when design-for-use meets design-in-use. *Interacting with Computers* 15(5): 647–663.

Rabardel, P. (1995). Les hommes et les technologies; approche cognitive des instruments contemporains. *Armand Colin* hal-01017462(1): 239.

Rabardel, P., & Beguin, P. (2005). Instrument mediated activity: From subject development to anthropocentric design. *Theoretical Issues in Ergonomics Science* 6(5): 429–461.

Shorrock, S. (2016). *The varieties of human work. Humanistic Systems: Understanding and Improving Human Work.* Available at: https://humanisticsystems.com/2016/12/05/the-varieties-of-human-work/ (Accessed 7 April 2022).

USDA. (2021). *Sugar annual. Sao Paulo ATO.* Report number: BR2021-0015. United States Department of Agriculture. Available at: Available at: https://usdabrazil.org.br/wp-content/uploads/2021/04/Sugar-Annual_Sao-Paulo-ATO_Brazil_04-15-2021.pdf

Vilela, R. A. G., Laat, E. F., Luz, V. G., Silva, A. J. N., & Takahashi, M. A. C. (2015). Pressão por produção e produção de riscos: a "maratona" perigosa do corte manual de cana-de-açúcar. *Revista Brasileira de Saúde Ocupacional* 40(131): 30–48.

19 New Scientific Methods and Old School Models in Ergonomic System Development

Thomas Hofmann, Deike Heßler,
Svenja Knothe, and Alicia Lampe
Hochschule Osnabrück

CONTENTS

I, Thomas, the lead author of this chapter, studied industrial design with a focus on ergonomic product and interface design. Since then, I have been primarily involved in the development and design of industrial products as well as safety-critical human-machine interfaces (HMI) in the field of software development. My team and I, who together wrote this chapter, design products for the business-to-business (B2B) market and transfer the practical and scientific findings from product design and ergonomics into teaching and research. I do this as part of my professorship at Osnabrück University of Applied Sciences with the goal to promote the constellation of practical development work, research, and teaching in a synergistic way.

Every day, we learn which findings from research and teaching are useful for product development and vice versa, meaning how practical experience supports

DOI: 10.1201/9781003349976-19

science. We often realise that many theoretical findings, methods, and models have only limited relevance for practice. This is what we cover in this chapter with a focus on ergonomic and usability aspects in product design to clearly illustrate the synergies and discrepancies between theory and practice. Where has the transfer of ergonomic knowledge into practice worked well and where has it not?

LET'S GET PHYSICAL: THE NEED OF HAPTIC MODELS IN ERGONOMIC DESIGN DEVELOPMENT

We were asked to design a new, compact, and extremely easy-to-maintain industrial 3D scanner for a company named GOM metrology GmbH.[1] This process was to be carried out in close cooperation with the engineering, marketing, and sales departments of the company. Although the project presented several complexities, as usual in product development, here we wanted to focus on the usage-related characteristics. These were important for us to optimise product usability. More specifically, the scanner had to be designed very much with ergonomics in mind due to its special proximity to the user and its high mobility. This means that special attention had to be paid to the following aspects:

- Visible orientation (front/rear)
- Recognisable grip areas
- Good handling for mounting/dismounting on tripods
- Ease of maintenance (changing filters, lenses)
- Connectivity (plug mounting)

We preferred a haptic and physically tangible design approach and decided to work largely with physically experiential models, mock-ups, and simulations. This approach seemed the most practical to us, even if it was unusual in the context of this cooperation. Up to that point, due to the high importance of constructive elements, we had worked almost exclusively in computer-aided design (CAD) programmes and the target versus performance comparisons were taking place in the virtual space. Indeed, we had previous experience with physical models and their communication. The main challenge was the lack of experience of GOM with this procedure as in previous projects they had worked without physical models. Nonetheless, due to the very good relationship with all parties involved during for more than 20 years of our cooperation, it seemed a worthwhile idea to implement our knowledge from teaching and research into this practical project.

What was going to happen? In the worst case, we would have left the project with the realisation that the previous approach with CAD was more suitable than the new one. Indeed, one might ask whether physical models should be built at all today (Greenberg et al., 2012). However, based on our experience in the field of product design, it has become clear that a virtual representation of a product is still not fully capable of being assessed in a comparable way to a physical artefact. However, this virtual versus physical product debate (Warnier & Verbruggen, 2014) will not be

---

[1] https://www.gom.com/en/

presented further in the context of this chapter. Another question could be whether we would have lost our reputation with the new approach, should the latter fail. The answer is that we were in a 'safe to fail' environment. Underpinned by its philosophy, GOM is mostly open to new ideas and ultimately thrives on trying out new things.

Admittedly, it was and still is quite a comfortable situation to be allowed to try something new with a project partner. We have been fortunate to deal with project partners who accept novel approaches. Apart from the organisational appetite for innovation, we believe that this has to do with trust in us, the design service provider, but also perhaps with the fact that, as an institute at a university, we should and must, by definition, always try out new approaches to generate new input. From our point of view, this is a foundation of freedom in research and teaching, but perhaps also a stroke of luck at our university, which actively supports this principle (Spitz et al., 2021).

THE APPROACH

It was important for us to work on the correct scale from the beginning, in terms of both the scanner's external dimensions and its weight. After a short, basic sketching phase, we moved on to model building, which was carried out physically and in CAD. Relatively early, it was decided that the final product should not have a very expressive shape. This means it should simple enough with low-complexity geometry and easy to describe by an external non-designer (Figure 19.1). After all, it was an industrial product, which should follow the mantra 'form follows function'. Hence, it was soon clear the product would end up being a small black box. However, this did

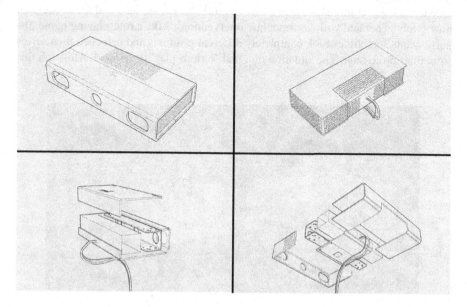

FIGURE 19.1 First conceptional sketches in CAD.

not necessarily make the design any easier. The product should have a simple shell design and be easy to maintain. At the same time, the case should also be robust and have clear signalling functions (e.g. Where is the front and the back? How do I open the unit without damaging it?).

PHASE 1

We started organising workshops in which both designers and non-designers (e.g. engineers, marketing experts, and senior management) could contribute to the design of the product in a low-threshold way. This means that our aim was not to find a final form but to collect initial ideas for the design and create the design framework from the point of view of all participants. We were convinced that workshops with active participation would make more sense than purely theoretical question-answer sessions.

In the first workshops, we brought with us many polystyrene foam blocks with dimensions based on the expected technical and functional components (power supply, light unity, inner construction parts, sockets, etc.). We used these blocks to draw meaningful parting lines, signalling functions, edges, etc., and then work on them with cutters and sandpaper. During the workshop, we asked participants of mixed expertise to get together in pairs and generate ideas. They were asked to design the foam blocks according to these ideas by using any of the available tools. More than 20 models were developed (Figure 19.2).

The participants initially seemed reserved to 'do design without being a designer'. However, after a few moments of initial scepticism, the participants identified with the method and were motivated to create a functionally and aesthetically interesting design. The participants carved and sketched their designs with childlike joy. Due to this exuberant mood, previously unthinkable discussions and ideas could be now safely generated with no academic interventions. Like a role-playing game, the participants sometimes took completely different positions and detached themselves from their profession. The situation of a real 'serious play' developed. Although this

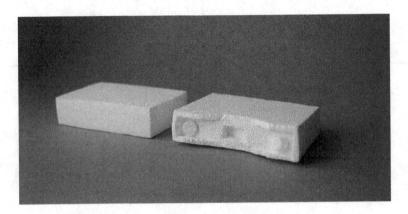

FIGURE 19.2   Basic foam mock-ups for the workshops.

created a huge mess of leftover materials, this approach had several highly positive effects on the project:

- The basic ergonomics and usability of the sensor (i.e. dimensions and measurements) could be rudimentarily assessed and discussed at this early stage.
- Many variants could be sketched and discussed within a very short time without having to work in CAD.
- The reception by all participants was very good.
- There were no problems in interpreting each model, which is a frequently encountered challenge in discussions based on CAD models.
- The integration of 'non-designers' into the design process was inexpensive and 'safe to fail'. There were no inhibitions to modify a model, sketch on it, or criticise it.

In general, this methodology was inviting for participatory design. The motivation of the participants was high, and an extremely constructive and open working atmosphere was created. Equally important, several usability characteristics were already worked out in this first phase. Admittedly, this method also worked well in this case because the product to be designed was of a graspable size (Figures 19.3 and 19.4). This greatly facilitated haptic (i.e. touch) and functional examination of the models developed.

Phase 2

Based on the first workshop results, we continued the design process in CAD. The development benefited from the fact that the participants had previously created models, which now could be recognised as 3D models on the computer. This was essential because it is often difficult for non-designers to evaluate CAD models aesthetically and functionally because they cannot imagine them physically. The CAD models shown in this phase were largely based on the models created in the workshop, and accordingly, the participants quickly recognised their own ideas. What's more, the

FIGURE 19.3   First cardboard models.

FIGURE 19.4   Cardboard models including mock-ups of the technical package.

usual negative critiques were not forthcoming; the latter are, perhaps, a bad habit of typical German 'brainstorming', whereby one of the participants criticises the others' ideas rather than just letting them stand. Why exactly this typically 'German' behaviour did not occur here, we can only guess. The most plausible seems to be that all those involved have had a very constructive way of working together and knew from the past how good the results become when little criticism is expressed about new ideas. Therefore, the design could effectively progress further in this phase. The gap among disciplines was eliminated because the participatory process integrated all participants (e.g. designers, usability experts, and constructors) into the design process right from the beginning of the project.

Phase 3

After clarifying how the design should look in principle during the first two phases (how the enclosure separations should be implemented, what material the enclosure should be made of, etc.), we moved to the final shaping phase. In this decisive phase, it is crucial to be able to assess details precisely, recreate the ergonomics as realistically as possible, and understand the product. For this purpose, realistic computer-generated imagery (CGI) representations are usually used, or animations of the final product are generated (Figure 19.5).

The disadvantage of this approach, particularly in the development of hand-held products, is that a real evaluation of ergonomics is hardly possible because the reference to haptic reality is missing. Although it is already possible today to visualise quite real-looking models on the computer, the physics that are so important for an assessment (weight, real dimensions, intuitive interaction, etc.) are missing. Our experience in product design projects shows that users interpret a product in CAD or virtual reality differently than a physical model. Especially, people who do not

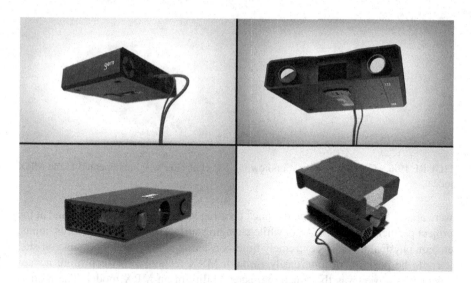

FIGURE 19.5   Collection of different design approaches for the final design phase.

constantly deal with CAD software or use other forms of visualisation (e.g. traditional clay modelling or foam-based shaping) try to interpret digital forms and bring in their own experiential knowledge. However, if you have a physical artefact in front of you, the brain can hardly create its own reality; the artefact is perceived by the senses exactly as it physically exists.

For this reason, we decided to use a mixed-reality approach (MRX). This usually means combining physical and virtual content into an overall context that is comprehensible to the viewer. This can mean mixing virtual content into a real space through augmented reality (AR) glasses or feeding a virtual car cockpit into virtual reality (VR) glasses while the viewer sits in a real car seat. In our case, however, we decided to take a different approach because our product was of a size that allowed to use a physical avatar enriched by a VR visualisation. The setup is explained in the following paragraphs.

We created a black box with geometric dimensions to the final product. The dummy, which was made of polystyrol foam (Figure 19.6), was fitted with weights to achieve the desired weight of the final product.

The virtual part of the construction consisted of several interchangeable CAD designs that were displayed in a VR headset. The viewer could change these designs independently. The different CAD models were precisely mapped onto the cardboard model (i.e. virtually projected onto it) with the support of a marker integrated into the physical model (VR tracker). This rendered the product tangible as the user could experience the different designs realistically. The surfaces and geometries could be visually grasped, and the physical weight was sensed.

The method above made it possible to test and validate about 25 different variants of the potential design within a very short time. This brought several advantages. Compared to the creation of several real models with different characteristics, we

FIGURE 19.6 Concept visualisation how the VR controller was implemented to the MRX model.

were able to save working time by quick creating variants in the workshop and at the project partner's premises. Also, while changing surfaces (colours, structures, joints, etc.) on physical models is extremely time-consuming and often not possible on-site, it could be performed quite easily through MRX technology. Much more advantageous, however, was the functional accessibility of an MRX model. The models could not only be moved and viewed but they could also be dismantled, opened, and disassembled. Thus, not only aesthetic aspects could be validated, but also ergonomic and technically functional parameters.

The acceptance of this method by the persons involved was very high, which naturally increased the efficiency of the procedure. In the beginning, there were slight difficulties in adapting the VR glasses and interacting with the dummy model. However, due to the highly immersive and intuitive interaction, the procedure was understood and followed within 1–2 minutes of demonstration (Figures 19.7 and 19.8). What further increased the enthusiasm for this approach was that the combination of VR visualisation and interaction with a physical avatar was completely unknown and thus fascinated everybody. Of course, the fact that the technology worked immediately also helped here! During this third phase, the basic design was defined and relevant decisions about the details were made.

PHASE 4

In the final development phase, we produced physical prototypes through 3D printing, followed directly by the pilot series. Due to the MRX simulations carried out in phase 3, in addition to the aesthetics and technical integration, the ergonomics and usability could be well examined, and almost all potential weak points could be identified and eliminated. Both the 3D-printed prototypes (Figure 19.9) and the pilot series were equipped with the technology components and assessed again for usability. The points of discussion identified during the MRX were confirmed and could be quickly remedied.

OUR REFLECTIONS AND NEXT STEPS

Although we had previously tested the methods described in phases 2 and 3 above in university teaching and identified them as an interesting approach to assessing design

FIGURE 19.7 The MRX visualisation the user sees via the VR glasses.

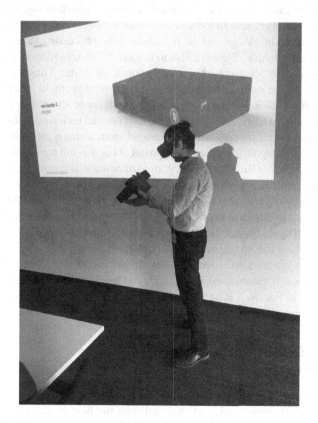

FIGURE 19.8 Live demo of the MRX experience.

FIGURE 19.9   Production models with some marked details to be refined.

and usability, we had never conducted these with an industrial partner. Thus, in a way, we took a risk as to whether the approach would bring the desired acceptance and, above all, results. On the other hand, as stated above, GOM was a very open-minded project partner with whom we had built trust over the years and who was always receptive to new approaches in finding solutions. So, the initial conditions were favourable.

Basically, by using these methods we gained many important insights into practical parameters and were able to somewhat shorten the design process. However, apart from the technical knowledge and the reduction of the processing time, other aspects were much more valuable. We were able to involve all stakeholders of product development in the design process within a very short time frame. Thus, engineers, marketing experts, and sales agents were able to participate directly in the design process and they had fun with it. We also experienced that those new methods for ergonomic design could lead to a high level of identification with, and enthusiasm for, the process. In this way, the design process was transformed from a strict, mono-discipline process into an integrating task for the development. This was not achieved by a classical top-down approach, but through the integration of new participatory tools, which were developed in teaching. Therefore, we learned that tools used to motivate students in the class or laboratory can also work in the industry. We are glad that this experiment worked very well in this case and is increasingly being used by other project partners.

We are working on further expanding our competence in the MRX field, transferring design tools to the virtual world and porting to VR the phase using model construction foams. We are also expecting a few 'aha moments' from our project partners since this method is established in research and teaching but remains unknown in industry practice. For us, it is an ideal situation to be able to try out new approaches with students first and then use them with clients.

USABILITY OF STANDARD METHODS IN INDUSTRIAL DESIGN PROCESSES

In many industrial applications, work processes are becoming increasingly complex and are controlled, monitored, and analysed with the help of software. The task of user-centred design is to present this wealth of information for the users in a way that allows them to quickly detect, evaluate, and act upon this information. A customised

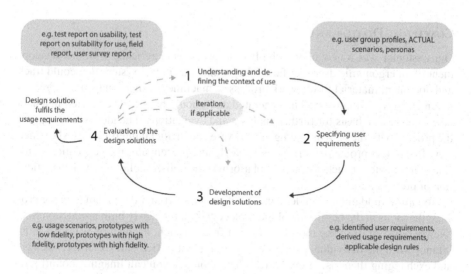

DESIGN PROCESS ACCORDING DIN EN ISO 9241-210

e.g. test report on usability, test report on suitability for use, field report, user survey report

e.g. user group profiles, ACTUAL scenarios, personas

1 Understanding and defining the context of use

Design solution fulfils the usage requirements

iteration, if applicable

4 Evaluation of the design solutions

2 Specifying user requirements

3 Development of design solutions

e.g. usage scenarios, prototypes with low fidelity, prototypes with high fidelity, prototypes with high fidelity.

e.g. identified user requirements, derived usage requirements, applicable design rules

FIGURE 19.10 Design process based on DIN EN ISO 9241-210 (ISO, 2019).

interface can increase efficiency, employee satisfaction, and support error prevention (ISO, 2018). According to DIN ISO 9241 standard on the ergonomics of human-system interaction (ISO, 2019), the process for designing a user-centred interface consists of four successive but distinct phases with possible iteration loops (Figure 19.10).

- Phase 1: This phase serves to determine the concrete framework conditions for the project. For this purpose, all relevant user groups and stakeholders and their relationships and goals must be identified.
- Phase 2: It involves specifying the user requirements and therefore defining the exact features of the system. This considers not only user wishes but also the economic targets of the system, which can also influence work processes and organisational structures.
- Phase 3: During this phase, concepts for the design ideas are derived and developed from the analysis of the requirements. The various results should be visualised (e.g. mock-ups, simulations, and prototypes) and presented to the users for evaluation to refine and concretise the design concept.
- Phase 4: The last phase, evaluation, is a recurring part of the process. Through evaluation, information can be gathered on new requirements, the strengths, and weaknesses of the design and the comparison with the boundary conditions can be made. The validation can be carried out either by the user himself or by an expert.

Depending on the complexity of the project, it can be difficult to obtain information or to ensure communication among participants with user-centred design of an interface. In our case, which is about the development of a higher-level control system for a production line in the automotive sector of a large industrial company, we describe some of these complications and suggest possible solutions.

Conceptual Development

Our customer was a new client who had no previous experience with the development of an ergonomic user interface. They wanted a control system that could track and document material flows, work processes, machine settings, and similar process parameters in their entirety. The generated data should have been collected in a data lake to serve as a basis for further analyses and evaluations. The idea was to support the process with the help of intelligent software and artificial intelligence (AI), which can often detect production errors more easily and present alternative solution steps. The prerequisite for such an analysis platform is the holistic collection and consolidation of process data.

To gain a fundamental understanding of the product and the manufacturing process, the focus at the beginning of the project was on the newly built production line. 'Newly built' in this case meant developed from scratch. Production was still in its infancy, with the individual departments working with Microsoft Excel spreadsheets and exchanging them using USB sticks. Therefore, as you can imagine, not all processes were smoothly coordinated at the beginning. Repeatedly, this became apparent during this project because we identified contradictions, insufficient agreements, or even unclear responsibilities among personnel from several departments.

The plan was to design the interface according to DIN ISO 9241 (ISO, 2018, 2019). Often, customers bring their ideas about how the project should proceed, including the generation of intermediate results and even the working methods and role of the designer. An indication of such an intention on our customer's side could be seen during their provision of documents and information. The customer was initially unaware that the work processes on the respective machines should form the basis for the concept's creation. Therefore, process diagrams and flowcharts were only handed out gradually instead of from the start of the project.

Furthermore, the customer believed that only a 'makeover' of the already existing forms for data input was sufficient, and no major research was necessary. Our first lesson learned was that we must always keep in mind that the client might not have the same background and knowledge regarding user-centred design as the designer. When first cooperating with a new project partner, this can quickly become an unpleasant trap. If the expectations differ greatly, misunderstandings can arise, which, in turn, may lead to resentment and frustration in the collaboration. Retrospectively, we realised that we had failed to explain to the customer that a higher-level structure could be devised and described only through a thorough understanding of the processes and the associated information by collecting input from the users. Accordingly, the initial phase was quite lengthy and cost a grey hair or two on both sides. From the beginning, it became clear that the application of the DIN ISO 9241 standard to the project would be difficult.

In addition, we had to adapt our working methods to the circumstances of the COVID-19 pandemic. This meant creating new methods for us and changing established procedures. Due to the pandemic, we could learn about the interactions and interdependencies of processes only by using documents prepared in advance by the customer and distilling information during online meetings. The lack of face-to-face contact delayed the establishment of trustful communication because there were only a few opportunities for small talk between the scheduled remote, web-based appointments. All participants were heavily consumed by their daily business, which limited the time for joint discussion sessions.

Apart from personal contact, the observation of the actions on site can be more enlightening than just a description, especially in the case of routine work. Everybody who tries to describe routine work will mostly notice that several small intermediate steps are missed from such descriptions; however, it is often exactly these small steps that could make the difference. In this context, we would like to mention an 'experiment' we conduct during our teaching and research. We always ask students in the introductory ergonomics course to think of all relevant steps for everyday activities (e.g. make coffee, water flowers, or shave) and design simple instruction manuals. Indeed, it happens again and again that essential intermediate steps are not identified. When asked why the protective cap on the razor was not removed before shaving, the answer is usually 'Well, that goes without saying...' - a typical misconception. NO, it cannot be taken for granted if you have never done this procedure before. This is an important insight that we were able to transport from our teaching into our practical project works.

Hence, at that point of the project, we once more realised that an on-site visit to the company could produce far more value and insights than a telephone conference. Nonetheless, we believe that the high motivation of the employees and their willingness to try out new methods could compensate for the omission of the on-site visits to a large extent. The first attempt to collect general information with the help of questionnaires got off to a slow start because the production process was simply too complicated and interwoven to explain in simple sentences. In contrast, switching to interviews was a complete success.

We asked the customer to collect ideas latently. It turned out that some of the users had already thought about the requirements for the new system in advance without our intervention, and they had written them down. By giving the users the opportunity to communicate any concerns, wishes, and individual ideas, they realised that they could actively participate in the design and their input was crucial to the success of the project. Consequently, the motivation of the employees to be involved in the further phases was very positive because they recognised the benefits for them as end-users. Building further onto this opportunity, we reacted flexibly to the needs of the users, which meant that any initial misunderstandings and knowledge gaps on both sides could be eliminated.

CONCEPT DEVELOPMENT

Theoretically, according to the method described in DIN EN ISO 9241-210 (ISO, 2019), the definition of requirements follows the research phase. We had observed this was feasible in fictitious or student projects where usually no unexpected external occurrences take place. However, this strict separation and processing of the individual phases might not be always possible in real-world conditions. In our project, the information was collected throughout the first phase and was immediately linked, analysed, and checked. As a result, no important information was lost, and all findings could be related to each other. Likewise, additions could easily be included in the research and requirements. When visualising contexts and processes from the research phase and the interviews, the creation of flowcharts, screens, and mock-ups proved to be extremely helpful.

Moreover, the workflows on the machines were so extensive that a separate flow diagram was created for each work area. This made it easy for users to check whether the sequences and the relevant information were reproduced completely and correctly. Allowed by the detailed collection of information, overlapping structures were quickly recognisable and could be easily included and visually prepared in the conceptual design. It also became quickly clear that the production line could not be viewed as a stand-alone entity. Instead, other departments such as warehouse logistics and quality management had a considerable influence on the way that the interface worked and was designed. However, this expansion of the user group meant that further interviews had to be conducted, leading to a longer and more extensive analysis phase. The challenge here was that the new interviews could not be performed with the acquired routine because these persons had not been involved in the first process phase and, thus, had no prior knowledge of the interview procedure. Therefore, the new user groups had to be informed about the procedure, which led to delays in the design process and to a greater amount of work on our side.

Moreover, in the form of an interesting and insightful by-product of the interviews, it became apparent that some designations and responsibilities had not been established across departments. This communication deficit between the individual departments had never been noticed before our cooperation. By uncovering, discussing, and eliminating these discrepancies, we were able to encourage the customer to define a uniform nomenclature and clear distribution of tasks in the control system. Hence, in such situations, the designer might also play the role of mediator and initiator, helping to clarify problems and disagreements in workflows and encouraging information exchange between different departments.

Iterative consultations with users refined the structure and usability of the interface. Every now and then, our customer also saw mock-ups of the interface and reacted with surprise to placeholder texts such as the names of materials and projects. In our experience, this is unfortunately not an unusual case with new collaborations, as the difference between a mock-up and a finished interface is not always understood. Thus, it was our task to clarify that in this phase the structure was in the foreground, and we would gladly consider the desired designations when this information was available. This scenario was repeated several times. Only after some time, everyone understood how to read the prototypes and first drafts. In this aspect, it becomes obvious it is sometimes a great challenge for a designer to design in a user-centred way when the user group is technically very deep in its own processes and more focused on details than seeing the big picture. Therefore, we had to point out the overall concept again and again and address technical specifications later in the fine-tuning work.

IMPLEMENTATION

The degree to which we as designers are involved in the implementation process varies from project to project. In this case, the implementation of the interface concept was to be managed by a third-party company. The communication with these developers was completely handled by our customer, which meant we were positioned outside this loop. The programmers of the developers had access only to our concept description and the style guide. Since we did not have access to the implementation status, we could not know the extent to which questions were answered correctly by the customer's contact person. Admittedly, this situation was extremely unsatisfying for us because it created the feeling that we had only completed half of the project.

Especially in a user-centred design, the phase of testing after implementation is the exciting one. This is where we find out whether the processes and requirements were analysed correctly, and whether the users are optimally supported in their work. Although the customer confirmed the users were satisfied with the interface, we are aware that there is always room for improvement. Weaknesses and misunderstandings can only be identified with the help of extensive feedback from the user and by observing user behaviour and adapting and optimising the respective processes accordingly.

THE AFTERMATHS

In conclusion, it was not possible for us to carry out the project in the sense of DIN ISO 9241, which, though, served us well as a guideline. Since new fields of work kept emerging during the project and its scope was completely misjudged by everyone involved at the beginning, it was almost impossible to meet the customer's request for a rough concept after just a few weeks. This case shows once again that an atmosphere of good communication is the basis for successful cooperation. Even if it is sometimes tiresome to explain things repeatedly, it is important to maintain a respectful and friendly communication style, even in difficult situations.

Not only the communication between contractors and clients but also in-house communication can lead to tensions. The task of the designer is then to act as a translator to clarify the misunderstandings. Especially in projects like this one, with a large interdisciplinary team from different departments, finding a 'common language' is essential. In our case, it was not clear for a very long time who the actual project sponsor was and what authorisations existed for this person to significantly control and influence processes. Each department spoke about the sponsor, but when asked, each department meant a different person! Only an interdepartmental meeting could finally clarify who this ominous sponsor was.

Furthermore, our experience from this project showed that the various disciplines do not only differ in the ways of communication but also their approaches to problems. Engineers like to fall back on proven and established solutions when solving problems. We, the designers, on the other hand, consider many possibilities, even seemingly crazy ideas, to reach the goal and only later do we check several possibilities and combinations to ensure feasibility and usefulness.

Finally, it must be pointed out that, although we would have liked to accompany the implementation, we can still look back on a successful project. The customer has gained great confidence in our work and would like to carry out an extension of the system with us. Thus, even where the practice deviated from the theory, we were able to find alternative solutions and compromises together with the help of the team. We are curious to see what will await us in the follow-up assignment.

REFERENCES

Greenberg, S., Carpendale, S., Marquardt, N., & Buxton, B. (2012). *Sketching User Experience, MK Morgan Kaufmann*, Elsevier, Amsterdam, The Netherlands.

ISO. (2018). *DIN EN ISO 9241-11: Ergonomics of Human-System Interaction - Part 11: Usability: Definitions and Concepts*, International Standardisation Organisation, Geneva, Switzerland.

ISO. (2019). *DIN EN ISO 9241-210: Ergonomics of Human-System Interaction - Part 210: Human-Centred Design of Interactive Systems*, International Standardisation Organisation, Geneva, Switzerland.

Spitz, R., Böninger, C., Frenkler, F., & Schmidhuber, S. (2021). *Designing Design Education*, AV Edition, Stuttgart, Germany.

Warnier, C., & Verbruggen, D. (2014). *Dinge Drucken – Wie 3D Drucken das Design verändert*, Gestalten, Berlin, Germany.

20 It's Only a Reporting Form

Brian Thoroman
Queensland Rail

CONTENTS

Before the principles of effective human-centred, user-centred, or good work design were clearly articulated and communicated (Safe Work Australia 2015; Horberry et al., 2019), safety practitioners often learnt similar lessons through trial and error while implementing various safety interventions. While this may not have been the most efficient way to support safety in our organisations, it provided practitioners with on-the-ground experiences of what worked and what didn't in practice. Gratefully, we now have the language to describe and discuss these approaches to safety change management and can apply them to provide more robust and effective interventions.

The context of this case is the implementation of a new incident reporting from within the led outdoor activity sector. The led outdoor activity sector delivers facilitated or instructed activities in outdoor education and recreation. Outdoor education, in Australia, is an experience common to students from a young age. These outdoor education programmes occur over time frames ranging from a single session to multiple weeks and include activities such as bushwalking, canoeing, rock climbing, and cycling. Further, these programmes occur in a variety of settings, from purpose-built camps to public lands and across variable terrains ranging from urban environments to remote outdoor locations. Environmental conditions in which outdoor educations programmes take place can substantially affect the potential hazards encountered, including high winds and the potential for tree or limb fall, high temperatures and the risks of hyperthermia, and rainfalls leading to swollen rivers and flooded campsites.

In 2008, during a gorge-walking activity in New Zealand, fast-rising water from heavy rains created conditions in which six students and a teacher drowned (Brookes et al., 2009). The unpredictable and dynamic nature of outdoor education programmes means there is tremendous uncertainty around which decisions are made; this creates conditions in which injury and death can occur. Even in relatively contained programmes, such as centre-based camps at single locations, this uncertainty can lead to

DOI: 10.1201/9781003349976-20

fatalities, such as the drowning of a student during a dam swimming activity (White, 2014). The complex interactions among the variability of activities, locations, conditions, equipment, and people in a largely unregulated industry require that organisations delivering outdoor education must learn as much as possible from every aspect of their work, from routine work to near misses and incidents.

FAILING TO LEAD, LEADING TO FAILURE

To paraphrase a comment attributed to Dr Kerr L. White, 'Good judgment comes from experience which comes from bad judgment' (Farley, 2013). This has been true in my practice, and I think applies to most organisations. While this sounds like a tongue-in-cheek description of the learning process, it aligns well with some of the concepts of organisational learning such as double-loop learning and modification of decision-making or goals based on experience (Argyris, 1977). Throughout my career, applying human factors and ergonomics (HFE) in the domains of risk management and safety, I have seen and experienced this process several times.

Years ago, when still a novice practitioner, ripe for bad judgment and the associated learning, I was tasked with implementing a new incident reporting form for field use. At this point, the sector was primarily focused on addressing safety concerns from a people, equipment, and environment perspective (Dallat et al., 2017). The existing internal incident reporting process was, somewhat naturally, aligned with the prevalent approach that was considered best practice at the time in the sector.

During that period, the primary focus of the outdoor sector from a safety perspective was focussed on the development of skills, training, and experience for front-line field staff. Incident data were captured to identify what had happened with the expectation that the expertise in the field would then manage those issues locally, based on the abilities and competencies of individuals (Carden et al., 2017). While it is now becoming accepted that incidents in the led outdoor activity sector are better understood as system events (Salmon et al., 2010), this was not the case at the time. The effects of this focus on individual causal factors for incidents cascaded throughout the safety management system of the organisation.

The extant incident reporting form had been implemented for several years and was deeply integrated into the organisation at multiple levels. The reporting form was a key component of front-line worker training and in-field incident and emergency management processes. Accordingly, all organisation reporting and associated processes had been predicated on the data captured from this form for years. This included all the supporting administration processes and information technology as well as the management reporting processes such as quarterly and annual CEO and board reports. Nevertheless, the edict had come down from the executive that the organisation had committed to a new incident reporting method and documentation form. In hindsight, this top-down mandate should have been a red flag for the reasons I explain below. Back then, it was not perceived as such.

Changing a form is one thing, but shifting thinking is another. All our organisational, technological, and human systems were geared towards the previous incident form and coupled with a strong cultural belief that we, as an organisation, were already sector-leading in our approaches. Hence, resistance to this change wasn't futile, it was inevitable. I didn't know it at the time, but there was already reticence across the teams with a sense of distrust with the new incident reporting form coming from an external research group. In my blissful ignorance, I simply ploughed forward.

The word from the executive was to 'just get it out there' because 'it's only a reporting form'. Therefore, I began putting the new form into the hands of front-line operational staff. They were given no training and no communication or consultation, and none of the administrative staff was informed of the change. I believe you can guess what the response was: loud and consistent refusal, grumblings from front-line staff, loss of safety data, complaints the new form was considered overly onerous and inapplicable to the types of incidents found in the sector, and on and on.

Reflecting now on the rollout of the new safety intervention, and with the benefit of hindsight, I can identify what went well and what went poorly from this initial attempt. What went well is that the project had executive leadership support, a key component that is often missing from safety initiatives and is critical to implementing organisational change (Cameron & Green, 2019). Another component that went well was the intent of the incident form itself.

The incident form had been developed by safety science academic researchers and was consistent with the modern systems thinking approach to safety. It was designed to capture contributory factors across the socio-technical system rather than those only at the so-called sharp end of work. This was done using a contributory factors framework to identify the impacts of decisions and actions from not just front-line workers but also those arising from the 'blunt end', including company management, regulators, and government. This exact intent of the project to improve safety for our organisation and our sector was one of the drivers of the adoption of the incident form.

However, while front-line workers appreciated the intent of the initiative to improve safety, the implementation of an effective approach to safety change management was sorely lacking. The bad news was that few, if any, of the principles of good work design (GWD; Karanikas et al., 2021) were applied (Table 20.1). Furthermore, the principles of human-centred (or user-centred) design (HCD; ISO 9241-210:2010) were also not considered, such as the design being driven and refined through user-centred evaluation or the design team including multidisciplinary skills and perspectives. Due to this lack of application of the principles of GWD, the initial attempt to deliver a modern incident analysis form to the front-line staff resulted in a complete failure.

Overall, the cultural conflict between how work was done and the vision for work in future was not identified at the time. Now, with the benefit of further education and experience, I have a better understanding of the so-called research and practice gap (Chung & Shorrock, 2011). At the time, I had no idea that such a concept existed, and I certainly did not understand how to address it to implement a modern HFE-informed safety intervention or apply effective principles of change management (Mento et al., 2002).

Following on from the failure of the initial rollout attempt, I had an opportunity to highlight the challenges of delivering such a top-down design approach and explain how the failure of the rollout was a system-wide issue that required a system-wide

TABLE 20.1

Evaluation of Initial Incident Form Rollout against Principles of Good Work Design (GWD; Karanikas et al. (2021) and Human-Centred Design (HCD; ISO 9241-210:2010)).

Component of GWD	Principles of HCD	Analysis of application in this case (failed phase)
Discovery – engage people	Users are involved throughout design and development	No end-user engagement.
Discovery – study context, tasks, work, and jobs	The design is based upon an explicit understanding of user, tasks, and environment	Context, tasks, work, and jobs were known and analysed; however, this information was not included in the rollout strategy.
Design – develop concepts	The design addresses the whole user experience	The incident form was delivered by a third party; no changes were allowed in the initial rollout. No end-user engagement in design.
Design – trial iterative prototypes	The process is iterative	
Design – determine acceptable trade-Offs		The consequences of the new incident form were not analysed. No goal conflicts, trade-offs, or variability were considered.
Realisation – business areas integration	The design team includes multidisciplinary skills and perspectives	No integration with other business areas. Information technology, training, and product delivery were not engaged in the initial rollout plan.
Realisation – consultation and implementation		No consultation prior to implementation
Realisation – evaluation and monitoring	The design is driven and refined by user-centred evaluation	No planned evaluation or monitoring.

solution. The loss of incident data and negative end-user feedback from every programme under the rollout was a clear indication that something had gone drastically wrong. Gratefully, the failure had been so significant, that there was executive leadership support to try another approach.

LEARNING FROM FAILURE

Our initial activity was to unpack all the ways in which the first attempt had failed, using a small learning teams' approach to understand stakeholder impact. Team members included (1) front-line operational staff, (2) the information technology teams who would need to be engaged with implementing the changes to the database, (3) the administrative staff who would be entering data into the system, (4) the researchers who had developed the original draft, and (5) the executive who would need to allocate resources to support the project. They were all brought in and encouraged to describe the potential impacts of the safety change from their point of view. From the perspective of GWD, we had begun the iterative process of engaging stakeholders to learn the shortcomings of the initial rollout.

Reassessing stakeholder engagement provided insights for the second phase of the discovery process to understand the context and breadth of the changes we were trying to implement. We realised that we had underestimated the impact of this project on all aspects of the organisation and would need to consider integration, implementation, and evaluation to better understand how stakeholders were impacted by this seemingly small change. All our incident trends and management reporting to internal stakeholders and board members were predicated on the existing form. Similarly, all safety interventions, trend analysis, and training systems that we used as part of our customer value proposition were based on the data that had been previously collected. Expectedly, there were widespread concerns that all that learning would be lost during the transition. Thus, our 'one simple change' created issues at almost every level of our system.

Analysis from that first rollout taught us several important lessons in relation to our incident form design and implementation. First, incident form design has a large effect on the quality, amount, and type of information captured. Second, good incident form design requires engagement with the first order front-line operational staff who will be using it as well as the second and third-order end-users who perform supporting tasks. Third, any design process must be iterative with rounds of piloting, feedback, and refinement. Finally, we realised that implementation requires a significant allocation of resources and commitment to collaboration across the levels of the organisation to ensure the success of the project.

MOVING FORWARD FROM FAILURE

Once we had begun to better understand the problem, we were able to take a much more user-centred approach to the design of the form. By considering the purposes of the form from both research and practice perspectives, user tasks and workload, including time pressures, as well as broader organisational requirements, we were able to come up with a draft form that would fulfil the needs of all our stakeholders. This draft was then reviewed by front-end users and modified to create a version that met the needs as we understood them and was usable. This process of structured end-user engagement created momentum and interest from our main user group in support of our in-field design trials.

We ran a series of short, iterative design pilots where we used a draft for a week or two with a select group, got feedback and then refined and re-trialled the form. Each iteration was reviewed by the end-users and the development team to ensure that changes did not negatively impact or misconstrue the key purposes of the form. Over time, the form became continually refined until a workable product had been produced. As this design phase was occurring over the course of several months, there was simultaneous engagement with the onset of the ongoing realisation process.

Due to the higher speed of the iterative design phase and the delay associated with large system changes, the process and timing of the back-end changes also had to be considered. Additional challenges included the additional complexity of providing data to the research group that had developed the original new incident reporting form while satisfying organisational reporting needs. In the end, it was determined to initially run the two systems, legacy and updated, in parallel until the updated system provided all the necessary functionality to replace the legacy system. To facilitate this transition to the realisation phase, back-end process teams were engaged.

In the early parts of the design phase, we began to communicate with back-end process teams (e.g. information technology, training, staffing, and administration) to understand operational requirements and build design briefs for necessary system changes (see Table 20.2 for a summary of project impacts across teams). Because these larger system changes take quite a lot of time and resources, it was crucial to involve these teams early in the process. This way they could understand the scope of the change requirements and do preliminary work and plan for future resourcing needs.

The initial parallel systems implementation phase was supported by the form design process itself. One of the updated design criteria from the discovery phase was to continue to capture the key organisational reporting data as well as the additional data in support of the research group. During this transitional phase, additional staffing resources were required to manually enter data into both systems. While this was the most expensive in terms of resource allocation, it provided the smoothest transition and ensured no loss of valuable safety data.

While the back-end technology and administrative processes were being developed, broader organisational processes and staff training were also being integrated. The process of integrating all business areas required several changes to the safety management system. Reviewing, consulting, and updating changes to the relevant policies and procedures required yet another cycle of discovery, design, and realisation across the organisation. Similarly, working with training and development to integrate the new form and approach into the staff training required its own cycle of

---

TABLE 20.2
Summary of the Project Impacts

Team	Summary of safety change impact
Information Technology	• Design and develop new database architecture • Design and develop new user interface • Transfer existing data in new system
Training	• Update and deliver new incident training module • Create new training qualification to track staff skill rollout • Manage staff qualification updates
Staffing	• Allocate or source human resources for data entry • Train human resources for translating new incident form into legacy system during parallel running
Administration	• Update safety management system • Provide assurance activities for data collection and data entry • Update and communicate changes to reporting • Provide assurance activities and analysis between legacy and updated systems during parallel running

TABLE 20.3

Actions Taken after the Failed Initial Rollout to Apply Principles of Good Work Design (GWD; Karanikas at al., 2021)

Component of GWD	Analysis of application in this case (successful phase)
Discovery – engage people	Multiple interviews and observations to better understand the impacts of the change on the front-line workers.
Discovery – study context, tasks, work, and jobs	Analysis of the work context and tasks competing for front-line worker resources that impacted the ability to use the incident form.
Design – develop concepts	End-user feedback and participatory design to develop incident form concepts.
Design – trial iterative prototypes	Short two-week trials with iterative design changes done with a core team of front-line workers acting as change champions.
Design – determine acceptable trade-Offs	Prototypes were evaluated on the extent to which they met the dual purposes of organisational reporting and research partner requirements.
Realisation – business areas integration	Training teams were involved in the development. Integration with information technology led to the additional requirement to run dual systems, and executive team allocated additional administrative resources to support short-term interim solution.
Realisation – consultation and implementation	Broad consultation across organisation led to a phased implementation approach while internal information technology systems changes could be scoped and developed.
Realisation – evaluation and monitoring	Regular evaluation meetings on progress of design project and monitoring of ongoing data collection both for organisational systems and research partners.

the GWD process. Table 20.3 provides a summary of how we applied GWD principles across the organisation during the safety change process.

While it took nearly a year to get there, in the end, we were able to achieve the requirement to implement the new incident reporting form successfully and roll it out to the organisation. Soon after completing this process, I moved on to my next professional challenge at another organisation. Fortunately, in the organisation mentioned in this case, the implementation process continued with ongoing integration and improvement. Moreover, I am happy to report that continual innovation became an integrated process for its incident reporting. This included implementing digital incident reporting and integration with the national incident database for the sector. By implementing the fundamental principles of safety change management coupled with GWD and HCD, the organisation was able to position itself for success then and in future.

LESSONS LEARNED

One of the key lessons during this safety change management process was that the application of effective design principles requires an understanding and engagement of the impacted system. A single safety intervention will often thread throughout every aspect of an organisational or management system. I have found that safety change management and work design require an understanding based on a systems thinking approach to be successful.

A second key lesson for me was that applying the principles of effective design is itself an iterative process, and design never really ends. Rather than applying the principles to just 'the project', we found ourselves applying them, or a subset of them, at every stage. Discovery and realisation were critical approaches that were adopted during the learning process from our original failure. As we discovered the magnitude of the impact of the safety change, each user group went through the design phases, which were tailored to their needs. From IT to training, an iterative cycle of discovery, design, and realisation was required to accomplish the sub-tasks for the success of the overall project goal.

Each stage within the larger safety change management project is an opportunity to apply the effective design principles of discovery and engagement, iterative design, and integrated implementation and realisation. It has been stated that the strongest leverage points for system change are to modify or transcend our underpinning paradigms from which our work systems arise (Meadows, 2008). Thought from this viewpoint, the design principles may support a new way of thinking about safety change management and, therefore, become an embedded part of the organisations' actions towards safety improvement.

RECOMMENDATIONS

At the time of this case study, neither I, the organisation, nor the sector were aware of the tools available from the HFE discipline to support the design and implementation of safety interventions. I now understand that when a new safety intervention is considered, it is useful to take a structured systems thinking approach that aligns with effective design principles and applies appropriate HFE methods. This includes understanding the problem, context, stakeholders impacted, and the effects of the proposed safety intervention on the system. It is crucial that the principles of effective design are applied practically using suitable methods throughout the safety intervention lifecycle.

Meadows (2008) cautions us to make sure that we do not find ourselves pushing very hard in the wrong direction when intervening in systems. Therefore, it is critical to be thorough in our understanding of what we are trying to accomplish and understand the 'why' of any safety intervention. In the initial incident form rollout case, there was no clear purpose understood by the various stakeholders. In the subsequent application of the process, this identification of the 'why' of the safety intervention was accomplished by taking a step back to understand the purpose, followed by an analysis of the probable interactions arising from the new intervention across the work system.

Applying effective design principles at the scoping stage through engaging stakeholders and understanding the proposed work context can minimise issues later in the design and implementation processes. This initial process, which Norman (2013) describes as the beginning part of the so-called double diamond design process (i.e. first 'design' the problem, then design the solution) can identify whether the correct or 'real' problem is being addressed by the proposed intervention. Further, this process collects insights from stakeholders across the system which provides data that can be analysed with any HFE method in the discovery and iterative design phases.

The insights from the scoping and project discovery phase inform the iterative design phase. Applying robust HFE analysis methods, or subsets of these methods, as appropriate, such as Cognitive Work Analysis (CWA) (Rasmussen et al., 1994) or the Event Analysis of Systemic Teamwork (EAST) (Stanton et al., 2008) can be used to better understand the systemic influences on the proposed intervention as well as model the impacts of the proposed intervention across the work system. Such tools can further be used to simulate the effects of various design options. In turn, the artefacts created from these methods in the discovery and design phases can be used to support and inform the realisation phase.

System analysis methods applied during design provides useful insights into the realisation phase. For example, if the EAST method (Stanton et al., 2008) was used during the discovery and design phases, the task and social networks created during those phases could provide insights into the key stakeholders required for successful delivery during the realisation phase. Had CWA (Rasmussen et al., 1994) been used during the discovery and design phases, the results of the Social Organisation and Cooperation Analysis (Vicente, 1999) could provide this information. Regardless of the choice of specific methods, fit-for-purpose methods application provide the 'how' to practically apply the 'what' of principles of effective design. In my experience, the use of systems thinking-based HFE theory and methods, aligned with effective design principles, driven by a real business need, provides the best results when implementing safety interventions (Figure 20.1).

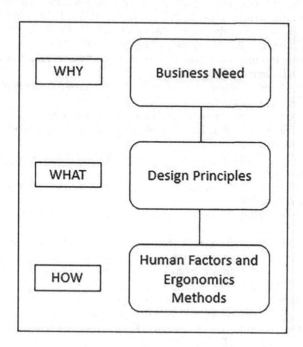

FIGURE 20.1 The relationship among business needs, design principles, and human factors and ergonomics methods.

REFERENCES

Argyris, C. (1977). *Double loop learning in organizations.* Harvard Business Review.

Brookes, A., Smith, M., & Corkill, B. (2009). Report to Trustees of the Sir Edmund Hillary Outdoor Pursuit Centre of New Zealand. Mangatepopo Gorge Incident, 15 April 2008. In *Turangi: OPC trust.* Retrieved from https://www.outdoored.com/sites/default/files/blog/files/091015-IRT-OPC_-Report.pdf (Accessed on Sept 6, 2022).

Cameron, E., & Green, M. (2019). *Making sense of change management: A complete guide to the models, tools and techniques of organizational change.* Kogan Page Publishers. London, England.

Carden, T., Goode, N., & Salmon, P. M. (2017). Not as simple as it looks: Led outdoor activities are complex sociotechnical systems. *Theoretical Issues in Ergonomics Science, 18*(4), 318–337. doi:10.1080/1463922X.2017.1278806

Chung, A. Z. Q., & Shorrock, S. T. (2011). The research-practice relationship in ergonomics and human factors – Surveying and bridging the gap. *Ergonomics, 54*(5), 413–429. doi :10.1080/00140139.2011.568636

Dallat, C., Goode, N., & Salmon, P. M. (2017). 'She'll be right'. Or will she? Practitioner perspectives on risk assessment for led outdoor activities in Australia. *Journal of Adventure Education and Outdoor Learning,* 1–17. doi:10.1080/14729679.2017.1377090

Farley, R. (2013). A is for aphorism–'Good judgment comes from experience; experience comes from bad judgment'. *Australian Family Physician, 42*(8), 587–588.

Horberry, T., Burgess-Limerick, R., Storey, N., Thomas, M., Ruschena, L., Cook, M., & Pettitt, C. (2019). An Introduction to User-Centred Safe Design. In *The core body of knowledge for generalist OHS professionals.* Australian Institute of Health and Safety. Tullamarine, Victoria. https://www.ohsbok.org.au/bok-chapters/

International Standards Organisation. (2010). Ergonomics of human-system interaction: Part 210: Human-centred design for interactive systems. ISO 9241-210: 2010.

Karanikas, N., Pazell, S., Wright, A., & Crawford, E. (2021). The What, Why and How of Good Work Design: The Perspective of the Human Factors and Ergonomics Society of Australia. In *International conference on applied human factors and ergonomics* (pp. 904–911). Springer. Cham.

Meadows, D. H. (2008). *Thinking in systems: A primer.* Chelsea Green Publishing. White River Junction, Vermont.

Mento, A., Jones, R., & Dirndorfer, W. (2002). A change management process: Grounded in both theory and practice. *Journal of Change Management, 3*(1), 45–59. doi:10.1080/714042520

Norman, D. (2013). *The design of everyday things: Revised and expanded edition.* Basic Books. New York, NY.

Rasmussen, J., Pejtersen, A. M., & Goodstein, L. P. (1994). *Cognitive systems engineering.* Wiley. New York, NY.

Safe Work Australia. (2015). *Principles of good work design handbook.* Safe Work Australia. Canberra, Australia.

Salmon, P., Williamson, A., Lenne, M., Mitsopoulos-Rubens, E., & Rudin-Brown, C. M. (2010). Systems-based accident analysis in the led outdoor activity domain: Application and evaluation of a risk management framework. *Ergonomics, 53*(8), 927–939. doi:10.1 080/00140139.2010.489966

Stanton, N. A., Baber, C., & Harris, D. (2008). *Modelling command and control: Event analysis of systemic teamwork.* Ashgate. Aldershot, England.

Vicente, K. J. (1999). *Cognitive work analysis: Toward safe, productive, and healthy computer-based work.* CRC Press. Boca Raton, FL.

White, P. (2014). *Inquest into the death of Kyle William Vassil.* Coroners Court of Victoria. Southbank, Australia.

21 SAfER Way to Design Work

Maureen E. Hassall
University of Queensland

CONTENTS

Approximately 360,000 workers die each year from work-related injuries (World Health Organization & International Labour Organization, 2021). In many systems, workers are often relied upon to detect and respond to unsafe situations to prevent and/or mitigate accidents. If human endeavours are successfully supported by work systems, this can lead to fortuitous outcomes where safe, efficient, and effective operations are maintained. However, if high-hazard industries are not designed to support the humans responsible for the safety of these systems, then major accidents can occur. In addition to major disasters, many fatality events also continue to occur with work-related traffic crashes causing the highest number of fatalities in several jurisdictions.

Examples of major accidents attributed to flawed designs include the following:

- Chernobyl nuclear disaster: It occurred when the reactor became unstable because the control room operators did not have the necessary information to monitor where the reactor was operating with respect to its safe operating envelope (Det Norske Veritas, 2011).
- Kaprun disaster: Many people were killed because they were unable to evacuate from the burning train (Carvel & Marlair, 2015). The 12 survivors were guided to safety by a volunteer firefighter who knew to evacuate down, not up, the tunnel (National Geographic Channel, 2004).
- BP Texas City refinery explosion: This resulted from a tower being over-filled because the control room operators did not know the actual level in the tower. The hot hydrocarbon spouted like a geyser out of the top of a vessel and was ignited causing an explosion that killed 15, injured 180,

and severely damaged the refinery (U.S. Chemical Safety and Hazard
Investigation Board, 2007).

- Buncefield fuel terminal fire: It was also caused by overfilling. The operators were filling a tank with petrol but due to faulty gauges, they did not
know the actual level in the tank. As a result, the tank was overfilled, and
the spilt petrol formed a vapour cloud that exploded causing massive damage to the terminal and surrounding buildings (COMAH, 2011).

Historically, the causes of such incidents have been attributed, at least in part, to the
decisions and actions of people responsible for operating the hazardous systems. As
a result, significant effort has been invested to better understand such incidents to
determine how future reoccurrences can be prevented. This effort has included using
human factors approaches to design safer work systems. These approaches can be
normative, descriptive, or formative in nature. Normative approaches focus on how
the human and system should or is intended to behave. Examples of such approaches
include reviewing and refining operator manuals and procedures. Descriptive
approaches are based on how the human and system behave in practice. These types
of analyses can lead to recommendations about improving human knowledge and
skills through better training, procedures, communications, and supervision.

Formative approaches have been developed to analyse the range of ways work
could be done in a system and take the view that there is a broad range of ways
work can be performed within the system. Within safety-critical, hazardous industries, there are often options and variability associated with the way work is done.
For example, novices might approach a task differently from experts. Tasks done in
high-risk situations, such as a confined space or in hot/cold weather, might be done
differently to those done in low-risk situations. People under time pressure might
perform tasks differently than those who are not under time pressure. Also, when
performing work, different people and the same person on different days often make
decisions about efficiency versus thoroughness trade-offs that can affect task execution (Hollnagel, 2009). As a result, I selected to explore using a formative approach
to improve work design.

Formative approaches seek to understand the constraints and performance-shaping
factors that influence workers' selection and execution of work to identify system
design modifications that could help all workers maintain safe and effective operations across all system states. Cognitive work analysis (Rasmussen, Pejtersen, &
Goodstein, 1994; Vicente, 1999) and ecological interface design (Bennett & Flach,
2011; Burns & Hajdukiewicz, 2004; Flach, Monta, Tanabe, Vicente, & Rasmussen,
1998) are examples of formative analysis techniques. In the high-hazard industry
sector, applications of these approaches seem to focus on control room operations.
However, for safety-critical field and maintenance activities, there were limited analytical tools to help human factors and ergonomics professionals, or other interested
analysts investigate how to better support the broad range of worker responses that
might be adopted in a high-hazard industry.

I believed we needed an approach that offered insights into how to improve workplace designs and better support workers' decision-making, especially when they are
seeking to detect and manage safety-critical situations from both field and control

room contexts. My experience across mining, construction, metal manufacturing, and oil and gas industries highlighted that often workers were required to manually intervene to keep plants and machines operating and that system designs often did not prevent unsafe or facilitate safe human-system interactions when manual intervention was required. The engineering design seemed to be done without thought about human intervention especially in situations when the design did not operate normally.

As a qualified engineer, I was not sure how to address this gap, so I decided to study organisational psychology which was informative but did not provide any solutions. I then went on to do a PhD in cognitive systems engineering that explored "Methods and tools to help industry personnel identify and manage hazardous situations" (Hassall, 2013). The product of this research was the development and testing of the Strategies Analysis for Engineering Resilience (SAfER) technique. The theoretical basis for SAfER comes from the field of cognitive system engineering, specifically from cognitive work analysis, formative strategies analysis, and organisational resilience principles (Hassall & Sanderson, 2012).

The SAfER technique is based on the premise that workers' actions are underpinned by their situation assessment, choice, and execution of a response strategy (Figure 21.1). The factors underpinning situation assessments were derived from the work of Rasmussen et al. (1994) and Endsley (1988). The factors underpinning the response strategy was drawn from the work of Rasmussen (Rasmussen et al., 1994).

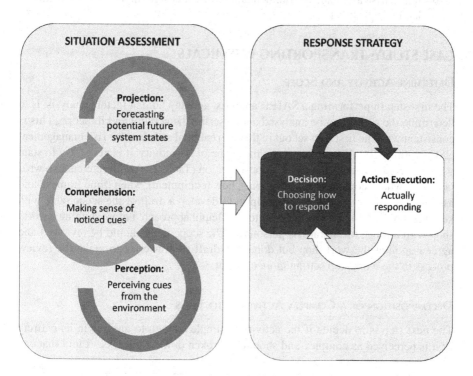

FIGURE 21.1 Factors underpinning worker responses.

SAfER analysis broadly occurs in two phases. First, the tool helps analysts identify the critical situation assessment indicators that signal safe versus unsafe operations. This allows analysts to identify design improvements which could make the indicators more salient and easier to comprehend. Second, SAfER uses categories to prompt the analysts to think about the range of strategies that workers might deploy to manage both normal and abnormal operating situations. The analyst then assesses whether these work strategies should be promoted, prevented, or tolerated. The outcome of the overall analysis can then be applied to the work design or other design interventions required to achieve the objective of promoting, preventing, or tolerating each strategy.

SAfER analysis can be performed on a conceptual system as a prospective risk assessment technique and during the investigation of actual incidents using the following process, which I explain and demonstrate through a case study in the next section:

1. Determine the activity to be analysed and the scope of the analysis.
2. If the activity is complex, decompose the activity into tasks using a Contextual Activity Diagram.
3. If the range of ways that the activity/task can be performed is not well understood, use the decision ladder framework to explore how the activity/task might be performed.
4. Complete the situation assessment analysis part of SAfER.
5. Complete the response strategies analysis part of SAfER.
6. Summarise and make recommendations about design improvements identified in steps 4 and 5.

CASE STUDY: TRANSPORTING CHEMICALS

DETERMINE ACTIVITY AND SCOPE

The first step in performing a SAfER analysis, and any human factors analysis, is to determine the activity to be analysed and describe the activity and its scope. This is consistent with the first step set out in the international standard for risk management (ISO 31000, 2018). When describing the scope of an activity, it is necessary to state what is included and excluded from consideration (Table 21.1) with reference to what (activity), who (people), where (locations), how (equipment), when (time frames), and assumptions. In practice, it can be helpful to develop a draft of the scope table with key stakeholders using a process akin to the Delphi approach[1] before running a workshop with a range of operations personnel. The scope table should be reviewed and agreed upon at the workshop but doing the draft as prework can make the review process in the workshop setting more efficient.

DECOMPOSITION OF A COMPLEX ACTIVITY INTO TASKS

The next step is to decide if the activity is simple enough to analyse in its entirety or it is perceived as complex and should be broken down into tasks. A tool that can

---

[1] https://www.rand.org/topics/delphi-method.html

TABLE 21.1

SAfER Analysis Scope Table for Loading, Transporting, and Unloading Hazardous Chemicals

Description	Included	Excluded
Activity: It should specify whether it includes normal and abnormal operations, maintenance, startup, shutdown, etc.	Loading truck with chemicals Driving truck from factory to warehouses Unloading truck	Manufacture and packaging of chemicals Truck maintenance
People: The persons who could be involved in activity.	Company dispatchers and drivers Client receivers	Other company and client personnel Members of public
Locations: Areas where the activity might occur.	Factory dispatch area Public roads Customer warehouses in Australia	Other factory areas Private roads Retail stores Residences
Equipment: Equipment and plant associated with activity.	Loading equipment Truck and its instrumentation Unloading equipment	Other plant and equipment on site, on roads or in warehouse
Timeframes: When the activity might occur (e.g. duration, time of the day/days of year, continuously or intermittently, within shift/across shifts).	Anytime throughout year Transport time to cover distance up to 4,000 km	
Other assumptions: Captures any other assumptions made (e.g. about context).	Compliance with road rules including max driving hours and hazardous goods requirements.	

help decompose activities into tasks is the Contextual Activity Template (Naikar, Moylan, & Pearce, 2006). The latter maps activity milestones against roles, or milestones against work locations, or roles against work locations to identify where tasks are performed. I found that deciding what categories to use to map activities is best done by trying all categories to see the decomposition results and then deciding which decomposition produces the best set of subtasks for further analysis. An example template shown tasks for transporting chemicals is shown in Figure 21.2.

Once the activity or tasks to be analysed have been identified, it can also be helpful to capture the interactions between people and between people and the system. Developing a control diagram similar to those used in System Theoretic Process and Analysis (STPA; see Leveson & Thomas, 2018 for details) can be used to highlight key actions and interactions. A simple example is shown in Figure 21.3 for a truck driver. The truck driver controls the truck through interfaces such as steering wheel, pedals, gear selector, and dashboard controls. These control actions of the truck driver are influenced by feedback from the truck interface that indicate speed, direction, and key indicators of truck status as well as the status of the driver, the alarms, and the camera information. The company dispatch can also send and

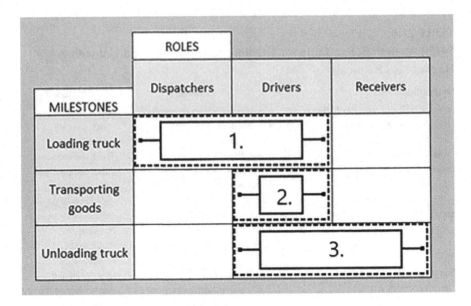

FIGURE 21.2 Example of a contextual activity template for loading, transporting, and unloading hazardous chemicals.

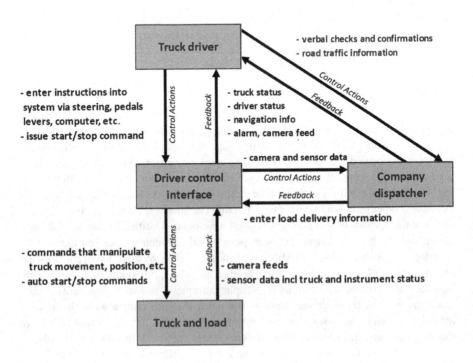

FIGURE 21.3 Example of control diagram.

receive information to and from the truck interface. Inputs into the truck interfaces can then be converted to actions that control truck movement.

DESCRIBING ACTIVITY/TASK ON DECISION LADDER

The next step in the SAfER analysis is to understand variabilities in the way that the work can be performed. Variability can arise from time pressures (e.g. rushed orders might be processed differently than low priority orders), risk levels (e.g. order involving highly volatile chemicals might be processed differently than an order involving inert chemicals), and task complexity. The task might be inherently complex (e.g. it might involve a lot of chemicals some of which might be potentially incompatible) or complex due to its novelty for the people performing it (e.g. because the task is new or because the people are new).

To understand the impact of any variability, it can be beneficial to describe these tasks using the decision ladder framework shown in Figure 21.4 (derived from Naikar et al., 2006; Rasmussen et al., 1994). The decision ladder framework shows the different cognitive activities (rectangles) and states of knowledge (ovals) that a person could use in performing a task. It also highlights that people can take shortcuts as shown by the arrows on the inside of the decision ladder. Testing of the decision ladder with industry people and novices highlighted that the decision ladder is difficult

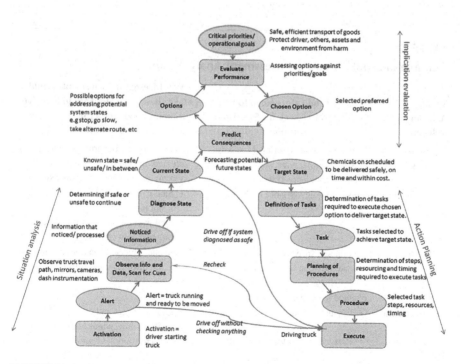

FIGURE 21.4 Example of decision ladder template. (Adapted from Naikar et al., 2006 and Rasmussen et al., 1994.)

to understand without help but could be usable and useful to promote in-depth thinking about how people could perform a task (Hassall & Sanderson, 2014). Based on this testing and subsequent work, I recommend that training on the decision ladder must be provided by skilled trainers so that it can then be implemented by users to prompt the discussions in the SAfER workshop.

SITUATION ASSESSMENT ANALYSIS

The SAfER table has two assessment sections: a situation assessment section and a response strategy section. The situation assessment section for the truck driver is shown in Table 21.2. Completing this table involves identifying the safety-critical indicators that will inform people of the system state and then determining how the design could make these indicators more salient, easy to perceive, comprehend, and project when required.

The second section of the SAfER table involves identifying the range of potential response strategies that could be deployed across a system, the possible consequences of these strategies, whether they should be prevented, tolerated, or promoted and how this can be achieved through design interventions. To assist with the identification of a range of possible responses, eight generic categories of strategies are provided. The categories are (Hassall & Sanderson, 2012; Hassall, Sanderson, & Cameron, 2014) as follows:

TABLE 21.2
Example of SAfER Situation Assessment Analysis Table

Situation Assessment Indicators	What Indicators Need to Be Monitored to Check for Safe/Unsafe Operation?	What Design Improvements Could Make These Indicates Easy to Perceive, Comprehend, and Project into the Future?
Equipment factors	Status of truck safety-critical systems: brakes, steering, tyres, speed control, collision avoidance systems.	Oral and text-based warnings with auto safe park feature when safety-critical systems are not functioning as required.
People factors	Status including vigilance of driver.	• Safe park system for truck if operator is not present. • Oral warning system for distracted, fatigued, unwell drivers.
Tasks factors	Location and speed of truck with respect to road: position in lane, clearance from other vehicles/obstacles, and actual vs planned travel path.	Actual and projected location information with oral directions and warnings.
Environmental factors	• Current and forecast weather • Presence of road/traffic disruptions	• Oral warnings and safe park when visibility low, winds high, etc. • Auto re-routing of truck around flood/fire impacted roads to minimise exposure to disruptions.

1. **Avoidance** category seeks to capture omitting, deferring, or forgetting to do a task. This would mean that some or all the processes identified on the control diagram and/or decision ladder would be skipped. Avoidance strategies can be preferable if it is not safe to continue but they can also produce adverse outcomes, for example, when an emergency response is urgently required.

2. **Intuitive** category seeks to capture the automatic or habitual response that involves first sensing and then executing without explicitly or deliberately using cognitive processes. Examples of intuitive responses can be derived from the decision ladder by looking to shortcut responses (e.g. those that go from activation or alert or observe information to procedure or execute). Intuitive responses are often used by experienced people. They can be beneficial if the intuition is correct but can produce adverse outcomes if situation is different that the intuitive assessment (e.g. a driver taking off at green traffic light without checking for oncoming traffic could be crashed into by a driver who goes through a red light).

3. **Arbitrary-choice** category seeks to capture responses that are guessed, scrambled, haphazard, or panicked. These types of responses can entail no consideration of options and include just random selection and/or execution of a response. Such strategies can be deployed in novel situations where there is time pressure and no knowledge or experience to inform alternative options.

4. **Imitation** category seeks to capture the responses that involve copying others or previously successful responses. Imitation strategies often involve shortcutting the definition and planning processes on the decision ladder. Imitation is often used to train people on the job and can be beneficial if the situation and task are identical to the ones imitated. If the situation and/or task differ, adverse outcomes might result.

5. **Option-based** category is when options are chosen without references to observed or noticed information. This can also involve selecting the first "good" option that comes to mind that meets the critical priorities or operational goals and not trying to find the "best" option. With respect to the decision ladder, this strategy can involve jumping straight to a chosen option by skipping the stages of observing information, diagnosing the state, predicting consequences, evaluating performance, etc.

6. **Cue-based** category seeks to capture those strategies that use observed information and predict consequences through cognitive processes to determine an appropriate response. These strategies involve using the steps associated with situation analysis on the decision ladder (i.e. the steps on the left leg of the framework) to identify the chosen option. The efficacy of cue-based strategies will depend on whether correct and valid cues were used.

7. **Compliance** category seeks to capture strategies that follow authorised rules, procedures, instructions, and guidelines. These types of strategies can be effective if guidance is correct and properly applied to the situation being addressed. For compliance strategies, the action planning steps of the

decision ladder (i.e. right leg) are executed in a manner that matches the authorised practices and processes.

8. **Analytical reasoning** category seeks to capture strategies that use analytical thinking or mental trial-and-error processes to reason the best way to perform the task. These types of strategies involve the implication evaluation steps of the decision ladder (i.e. top part). As such strategies can be time-consuming and cognitively demanding, they are best suited when there is available time and expertise to work through the problem.

To assist workshop participants in understanding the categories of strategies, I found it helpful to provide a simple example like travelling to work. With such an example, avoidance might refer to not going at all and working from home instead. Intuitive might refer to travelling to work using the usual mode (e.g. own car, ride share, or public transport), leaving at the usual time and using the usual route without thought. Arbitrary choice might involve randomly picking the mode, time, and/or route used perhaps because they deemed to be about the same. Copying the route or mode that someone else said would be quicker or easier is an example of an imitation strategy. Option-based category might involve taking the public transport bus because it just stopped outside the front door or going with a friend who was passing by. Checking traffic reports, public transport schedules, and/or condition of car before deciding is an example of a cue-based strategy. Compliance strategy could involve using the route recommended by the navigation system. Analytical reasoning could involve doing a detailed mental or spreadsheet type analysis to weigh up costs and benefits of using different modes, times, and routes before deciding.

The SAfER approach seems to elicit the most insights when brainstorming all the strategy categories for normal and abnormal situations. Shifts in strategies can occur with changes in time pressure, risk level, and/or other challenges when performing the task (Hassall & Sanderson, 2012). Using the truck driving example:

- The driver might begin driving by following the dispatcher's and navigation instructions to travel via roads and jurisdictions where hazardous chemical transport is permitted (compliance strategy).
- If the driver hears over the radio that the route has been closed due to a traffic accident, he/she might select another route previously used to still allow a timely delivery (imitation strategy).
- If this route turned out to be more congested than anticipated, the driver might then seek out information on alternate routes from the navigator and/or dispatcher (cue-based strategy).
- If whiteout blizzard/fog conditions turned out to be the cause of the difficulties, the driver might park the truck safety until conditions cleared (avoidance strategy).

An example of the response strategy section of the SAfER table for the truck driving case study is shown in the Appendix. Once the responses are identified, they can then be assessed in terms of possible reasons, implications, or impacts. The assessment should then inform the analyst's determination of whether the design should promote, prevent, or tolerate each response strategy. If a negative strategy cannot be prevented, then the design should tolerate it. This means the design should not allow adverse outcomes to result if the strategy is used. For example, if a human avoids performing a safety-critical function, then the design should have an automated response built in that will maintain safety.

The last step is to think of ways to make the design inherently safe and more user-centred to help workers select and execute strategies that lead to safe and resilient outcomes. This last step is crucial because, based on my experience, most if not all workers are trying to do the right thing at work and are not trying to cause an incident or accidents. The incidents or accidents occur when the work system is designed in a manner that induces or escalates unsafe situations. Hence, improving the design rather than blaming and retraining or recommunicating to the worker should produce a more sustainably safe system. I believe that "to err is human" so human error should not be considered a cause of an incident but a symptom of a work system that has not been fully designed for humans.

LESSONS LEARNED THROUGH APPLYING SAFER

I have applied SAfER to a variety of different case studies. Published examples include crane lifting task (Hassall, Sanderson, & Cameron, 2016), ship to shore petrol transfers (Worden et al., 2013), loading road tanker with toxic chemicals (Hassall, 2013), and work required of a librarian (Hassall, 2013). SAfER has also been applied to other industry scenarios, including human-autonomous mining equipment interaction scenarios, detecting, and addressing high-temperature hydrogen attack in processing plants and dealing with power outages in a petrol refinery. From these applications of SAfER, several lessons have been learned.

First, SAfER is best completed with a group of people knowledgeable of the activity being analysed. This produces a more complete range of safety-critical indicators, strategies, and design improvements. To ensure all the group members are involved, it can be helpful to ask each individual brainstorm their ideas first and then bring them into a group discussion to finalise the SAfER tables. However, as SAfER is a relatively new technique, training will need to be provided and expectations set as to the time requirements for a complete analysis. Teaching the scope, control diagram, and SAfER table is usually relatively straightforward, especially if an easily understood example case study is provided. Teaching the contextual activity template and decision ladder is more time-consuming and challenging because these concepts are often quite novel for people not exposed to human factors techniques. Instead of teaching these techniques, the persons leading the analyses might consider doing these steps themselves and then using the outcomes to guide and inform the group analyses.

Indeed, the brainstorming process often results in numerous responses for many of the categories. From experience, some categories, especially the option category, may not prompt any additional examples of responses. This is often because the actual response strategy has been captured under another category. This is the second lesson learned using SAfER: some strategies can be mapped to several categories. For instance, not doing something could be imitation, avoidance, or intuitive. It is not essential to define under which category the responses fall because the categories are just prompts to help people think of a more complete range of responses. Therefore, the most important is to capture as many safety-critical responses as possible.

It is also extremely helpful to explicitly state whether the strategy assessment relates to normal and abnormal operations. The importance of ensuring safe work design for abnormal operations (e.g. startup, process excursions, installation of temporary fixes, shutdowns) has emerged from the analysis of several accidents. For example, the Texas City refinery disaster occurred during start up (U.S. Chemical Safety and Hazard Investigation Board, 2007). The Flixborough disaster occurred after a temporary bypass was put in place (Health & Executive, 1975). The Chernobyl nuclear accident resulted from a process excursion that occurred during when the reactor was undergoing testing (International Atomic Energy Agency, 2011). The Xcel Energy accident was made worst due to inadequate emergency response (U.S. Chemical Safety and Hazard Investigation Board, 2010). In addition, I have analysed other incidents where an unrecognised change that triggered an abnormal situation such as a deviation from procedure has led to unsafe outcomes.

The third lesson learned is that SAfER was applicable to all case studies tried. It was even applied to concept systems, and the output could be used to inform implementation and risk management decisions. It was also applied as a risk assessment tool, and the output provided insights into design vulnerabilities and how the system could be made safer. Furthermore, it was applied to investigate an accident to determine how future accidents could be avoided. In this case, SAfER analysis was able to offer insights that would have been identified with traditional incident investigation approaches and additional insights into system design changes that should help humans deliver more successful outcomes across a range of different operating conditions beyond the actual accident scenario. However, it is important to note that, especially for complex activities, a good analysis can be time consuming and even tedious.

The fourth lesson learned is that SAfER focuses mainly on human activity and not technical malfunctions. Therefore, combining several technically focused approaches (e.g. HAZID, HAZOP, FMEA) with the SAfER technique can deliver more complete insights to improve the safety of sociotechnical systems. Indeed, several workshops have been run by using the same scope analysis and control diagram described in this chapter. Participants were asked to conduct a traditional HAZID, HAZOP, or FMEA analyses followed by a SAfER. The results suggested that the insights gained cover a more complete range of sociotechnical risks, and the recommendations from both the traditional techniques and SAfER are different but complementary.

In summary, SAfER can help identify ways to improve work designs by prompting the identification and thinking about the factors that promote good situation awareness of safety-critical indicators and prompt successful response strategies. It also aims to create systems that acknowledge and deal with the diverse ways humans can respond to both abnormal and normal operations. As such, SAfER differs from other human factors approaches that require humans to adopt an "one correct way" of performing tasks through the provision of procedures, training, rules, etc.

However, I have found that producing good SAfER analysis requires input from people knowledgeable about engineering good work systems and people knowledgeable about human-system interactions. I have also found that it takes effort to identify a succinct set of parameters to inform accurate situation awareness and translate strategies analysis into impactful design interventions. But if effort is invested, novel but practical insights into how to improve system safety through design (rather than just redoing training and procedures) can be obtained from SAfER. Simply put, SAfER can be a useful tool that complements other tools used to inform good work designs.

REFERENCES

Bennett, K. B., & Flach, J. M. (2011). *Display and Interface Design: Subtle Science, Exact Art.* Baton Rouge, LA: Taylor & Francis Group.

Burns, C. M., & Hajdukiewicz, J. R. (2004). *EID Ecological Interface Design.* Boca Raton, FL: CRC Press.

Carvel, R., & Marlair, G. (2015). A history of fire incidents in tunnels. In A. Beard & R. Carvel (Eds.), *Handbook of Tunnel Fire Safety* (pp. 3–24). London, England: ICE Publishing.

COMAH. (2011). *Buncefield: Why Did It Happen?* Retrieved from https://www.hse.gov.uk/comah/buncefield/buncefield-report.pdf

Det Norske Veritas. (2011). *Major Hazard Incidents - Arctic Offshore Drilling Review.* Retrieved from https://azslide.com/det-norske-veritas-major-hazard-incidents-arctic-offshore-drilling-review-nation_59eec8941723ddf436f3d50d.html

Endsley, M. R. (1988). Design and evaluation for situation awareness enhancement. In *Proceedings of the 23rd Annual Meeting of the Human Factors and Ergonomics Society* (pp. 97–101). Santa Monica, CA: Human Factors and Ergonomics Society.

Flach, J. M., Monta, K., Tanabe, F., Vicente, K. J., & Rasmussen, J. (1998). *An ecological approach to interface design.* Paper presented at the Human Factors and Ergonomics Society 42nd Annual Meeting.

Hassall, M. E. (2013). *Methods and tools to help industry personnel identify and manage hazardous situations.* (Doctor of Philosophy), The University of Queensland, Queensland, Australia.

Hassall, M. E., & Sanderson, P. M. (2012). A formative approach to the strategies analysis phase of cognitive work analysis. *Theoretical Issues in Ergonomics Science, 15*(3), 215–261. doi:10.1080/1463922X.2012.725781

Hassall, M. E., & Sanderson, P. M. (2014). Can the decision ladder framework help inform industry risk assessment processes? *Ergonomics Australia, 10*(3). Retrieved from https://espace.library.uq.edu.au/view/UQ:346420

Hassall, M. E., Sanderson, P. M., & Cameron, I. T. (2014). The development and testing of SAfER: A resilience-based human factors method. *Journal of Cognitive Engineering and Decision Making, 8*(2), 162–186. Retrieved from http://edm.sagepub.com/content/8/2/162.abstract

Hassall, M. E., Sanderson, P. M., & Cameron, I. T. (2016). Incident analysis: A case study comparison of traditional and SAfER methods. *Journal of Cognitive Engineering and Decision Making, 10*(2), 197–221. doi:10.1177/1555343416652749

Health and Safety Executive. (1975). *Flixborough (Nypro UK) Explosion 1st June 1974.* Retrieved from http://www.hse.gov.uk/comah/sragtech/caseflixboroug74.htm

Hollnagel, E. (2009). *The ETTO Principle.* Burlington, VT: Ashgate.

International Atomic Energy Agency (Producer). (July 10, 2011). *Frequently Asked Chernobyl Questions.* Retrieved from https://www.iaea.org/newscenter/focus/chernobyl/faqs

ISO 31000. (2018). *Risk Management - Principles and Guidelines.* Geneva, Switzerland: International Organization for Standardization.

Leveson, N. G., & Thomas, J. P. (2018). *STPA Handbook*. Retrieved from https://psas.scripts. mit.edu/home/get_file.php?name=STPA_handbook.pdf

Naikar, N., Moylan, A., & Pearce, B. (2006). Analysing activity in complex systems with cognitive work analysis: Concepts, guidelines and case study for control task analysis. *Theoretical Issues in Ergonomics Science, 7*(4), 371–394.

National Geographic Channel (Executive producer). (2004). *Fire on Ski Slope [Television series episode]. Seconds from disaster.*

Rasmussen, J., Pejtersen, A. M., & Goodstein, L. P. (1994). *Cognitive Systems Engineering*. New York, NY: Wiley.

U.S. Chemical Safety and Hazard Investigation Board. (2007). *Investigation Report - Refinery Explosion and Fire - BP, Texas City, March 23, 2005*. Retrieved from http://www.csb. gov/investigations/completed-investigations/

U.S. Chemical Safety and Hazard Investigation Board. (2010). *Investigation Report - Xcel Energy Hyrdroelectric Plant Penstock Fire - Cabin Creek Georgetown Colorado, October 2, 2007*. Retrieved from https://www.csb.gov/xcel-energy-company-hydroelectric-tunnel-fire/

Vicente, K. J. (1999). *Cognitive Work Analysis: Toward Safe, Productive, and Healthy Computer-Based Work*. Mahwah, NJ: Lawrence Erlbaum Associates.

Worden, J., Khanna, N., Campbell, J., Hassall, M. E., Sanderson, P. M., & Cameron, I. T. (2013). *An Integrated Approach to Technical and Human Factors in Risk Management of Ship-to-Shore Fuel Transfers*. Paper presented at the 43rd Annual Australasian Chemical and Process Engineering Conference (CHEMECA 2013), Brisbane Australia.

World Health Organization & International Labour Organization. (2021). *WHO/ILO Joint Estimates of the Work-Related Burden of Disease and Injury, 2000–2016*. Geneva. Retrieved from https://www.who.int/publications/i/item/9789240034945

APPENDIX

Example of the Response Strategy Section of the SAfER Table

What Plausible Strategies Could Be Used in the System Being Analysed?	Describe Reasons	Describe Possible Implications	Should Strategy Be Promoted, Prevented or Tolerated?	What Design Improvements Would Help Produce More Successful Outcomes?
Generic Strategy Prompt: Avoidance - Omit, Defer, or Forget to Do				
For normal operations: - Driver does not drive truck away - Dispatch does not give driver instructions	1. Loading not complete 2. Driver unavailable or unaware truck ready 3. Driver doesn't receive delivery instructions	Delayed delivery	1. Promote 2. Tolerate 3. Prevent	a. Driver receives real-time communications on truck loading status and dispatch/ delivery requirements
For abnormal operations: - Driver does not start or stops driving truck	4. Truck faulty 5. Traffic incident or road conditions means driver cannot proceed	Delayed delivery	4. Promote 5. Promote	b. Truck fitted with ABS, lane assist, proximity detection, collision avoidance, and speed control which are interlocked so truck cannot start or will initiate safe park when safety systems faulty or have not been serviced c. Driver's navigation instructions based on real-time traffic/road reports
For abnormal operations: - Driver does not deal with chemical leak/ignition	6. Driver unaware of state of chemicals	Potential fire, environmental damage and harm caused to humans by leaking/ignited chemicals	6. Prevent	d. Truck fitted with automatic load leak detection and collection and fire detection and suppression systems that alert driver and dispatch of unsafe load conditions
Generic Strategy Prompt: Intuitive - Automatic Response, Done without Explicitly, or Deliberately Using Cognitive Processes				
For normal operations: - Driver starts driving assuming everything is ok with driver, truck, load and travel route	7. Experienced driver and all previous occasions were ok 8. Driver in a rush or distracted	Driving while unfit to drive, with faulty truck, wrong load, and/or using incorrect travel route	7. Tolerate 8. Tolerate	e. Require driver to do prestart check on self, truck and load and have this confirmed by dispatch before truck can proceed (possibly with ignition interlocks)

(Continued)

APPENDIX (*Continued*)
Example of the Response Strategy Section of the SAfER Table

What Plausible Strategies Could Be Used in the System Being Analysed?	Describe Reasons	Describe Possible Implications	Should Strategy Be Promoted, Prevented or Tolerated?	What Design Improvements Would Help Produce More Successful Outcomes?
For abnormal operations: - Driver continues driving after receiving alarm or alert.	9. Info and alerts have been mostly false alerts in past 10. Driver distracted or fatigued/unwell	Unsafe operations that may result in a traffic incident or crash or leaking/ignition of chemical cargo	9. Prevent 10. Tolerate	As per c. and d. above f. Alarm/alert management system that minimises false alerts. g. System requires both driver and dispatcher to agree on alarm/alert response. h. Driver vigilance system that requires safe parking when driver cannot remain vigilant.
Generic Strategy Prompt: Arbitrary Choice - Guessed, Scrambled Haphazard or Panicked Response				
For normal operations: - Driver guesses which route, speed, clearances to use when driving truck.	11. Driver unfamiliar with route and/or truck	Could take longer, less safe route, could deliver to wrong location, could operate truck unsafely (e.g. speeding, getting it caught under bridges).	11. Prevent	i. Trucks navigation systems provides real-time oral and visual information to driver on best route, speed and lane to use (designed with b. and c.)
For abnormal operations: - Driver guesses what an alert/alarm means - Driver guesses how to deal with chemical cargo issues	12. Unclear or unfamiliar alert 13. Driver distracted or fatigued/unwell	Unsafe operations that may result in a traffic incident or crash	12. Prevent 13. Tolerate	As per h. above j. Simulation and emergency drill training

(*Continued*)

APPENDIX (*Continued*)
Example of the Response Strategy Section of the SAfER Table

What Plausible Strategies Could Be Used in the System Being Analysed?	Describe Reasons	Describe Possible Implications	Should Strategy Be Promoted, Prevented or Tolerated?	What Design Improvements Would Help Produce More Successful Outcomes?
Generic Strategy Prompt: Imitation - Copy How Others Do It or Copy What has Worked in the Past				
For normal operations: - Driver copies previously used routes and way of driving truck.	14. Experienced driver 15. Novice driver copying how he or she was shown	Might take longer, less safe route, could deliver to wrong location, could operate truck unsafely.	14. Tolerate 15. Tolerate	As per c. and i. above. k. Investigate truck differences and design options to mitigate potential adverse outcomes
For abnormal operations: - Driver copies previous responses to alarms - Driver copies previous responses to chemical cargo issues	16. Experienced driver. 17. Novice driver copying how he or she was shown *Implications: responses could be wrong leading to unsafe situations (e.g. truck or chemical incident).*		16. Prevent 17. Prevent	As per g. and d. above
Generic Strategy Prompt: Option-Based - Select Chosen Option from without Considering Observed Information				
For normal operations: - Driver selects route without referencing information provided	18. Rushed driver 19. Novice driver unaware of information provided	Wrong and potentially unsafe route	18. Prevent 19. Prevent	As per i. above
For abnormal operations: - Diverted driver selects route without checking - Driver selects alarm response without checking	20. Rushed driver 21. Novice driver	Wrong and potentially unsafe route	20. Prevent 21. Prevent	As per i. and g. above

(*Continued*)

APPENDIX (*Continued*)
Example of the Response Strategy Section of the SAfER Table

What Plausible Strategies Could Be Used in the System Being Analysed?	Describe Reasons	Describe Possible Implications	Should Strategy Be Promoted, Prevented or Tolerated?	What Design Improvements Would Help Produce More Successful Outcomes?
Generic Strategy Prompt: Cue-Based - Select Chosen Option Using the Observed Information/Cues and Predict				
For normal operations: - Driver closely monitors navigation systems and consults with dispatch	22. Driver sees benefit of relying on system	Optimum route used	22. Promote	As per i. and g. above l. Investigate system for measuring, providing feedback and rewarding good interactions
For abnormal operations: - Driver refers to safety sheets to know how to handle chemicals	23. Driver understands importance of correct chemical handling	Safety chemical issues	23. Promote	m. Investigate interactive system (e.g. like Siri) for informing driver of chemical cargo
Generic Strategy Prompt: Compliance - Following Procedures as They Are Written/Practiced				
For normal operations: - Driver follows road rules, operating procedure, and dispatch instructions	24. Driver understands importance of following procedures	Transportation in accordance with laws and company policies	24. Promote	As per j. and l. above
For abnormal operations: - Driver follows emergency response plan	25. Driver understands importance of following procedures	Chemical issues dealt with in accordance with laws and company policies	25. Promote	As per j. and l. above
Generic Strategy Prompt: Analytical Reasoning - Using Analytical Thinking to Reason Out the Best Way to Perform Task				
For normal operations: - Driver determines from own detailed analysis route to use	26. Driver does not trust system	Delayed delivery and/or wrong route	26. Tolerate	As per j. provide drivers who want to do this an opportunity to do so in the simulator

(*Continued*)

APPENDIX *(Continued)*
Example of the Response Strategy Section of the SAfER Table

What Plausible Strategies Could Be Used in the System Being Analysed?	Describe Reasons	Describe Possible Implications	Should Strategy Be Promoted, Prevented or Tolerated?	What Design Improvements Would Help Produce More Successful Outcomes?
For abnormal operations: - Driver thinks about and develops own emergency response	27. Driver unaware or chose to ignore available information	Delayed emergency response and incorrect actions	27. Tolerate	As per j. provide drivers who want to do this an opportunity to do so in the simulator

Index

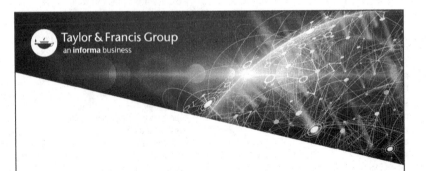

Printed in the United States
by Baker & Taylor Publisher Services